Palgrave Studies in Climate Resilient Societies

Series Editor
Robert C. Brears, Avonhead, New Zealand

The Palgrave Studies in Climate Resilient Societies series provides readers with an understanding of what the terms **resilience and climate resilient** societies mean; the best practices and lessons learnt from various governments, in both non-OECD and OECD countries, implementing climate resilience policies (in other words what is 'desirable' or 'undesirable' when building climate resilient societies); an understanding of what a resilient society potentially looks like; knowledge of when resilience building requires slow transitions or rapid transformations; and knowledge on how governments can create coherent, forward-looking and flexible policy innovations to build climate resilient societies that: support the conservation of ecosystems; promote the sustainable use of natural resources; encourage sustainable practices and management systems; develop resilient and inclusive communities; ensure economic growth; and protect health and livelihoods from climatic extremes.

Nyong Princely Awazi

Building Resilience

Climate Change and Livelihoods in the Global South

palgrave
macmillan

Nyong Princely Awazi
Department of Forestry and Wildlife Technology
University of Bamenda
Bambili, Cameroon

ISSN 2523-8124 ISSN 2523-8132 (electronic)
Palgrave Studies in Climate Resilient Societies
ISBN 978-3-031-93790-3 ISBN 978-3-031-93791-0 (eBook)
https://doi.org/10.1007/978-3-031-93791-0

This Palgrave Macmillan imprint is published by the registered company Springer Nature Switzerland AG
The registered company address is: Gewerbestrasse 11, 6330 Cham, Switzerland

Acknowledgments

The author is grateful to all those who have supported this book project in one way or another. I owe a debt of gratitude to the reviewers whose pertinent comments helped in fine-tuning and improving on the quality of the manuscript.

Contents

About the Author

Dr. Nyong Princely Awazi serves as Lecturer at the Department of Forestry and Wildlife Technology, College of Technology (COLTECH), The University of Bamenda, Cameroon. He holds a PhD in Agroforestry and Valuation of Ecosystem Services from the University of Dschang, Cameroon. Since 2014, he has been involved in research and consultancy on the cross-cutting themes of agroforestry, forestry, ecotourism, the bioeconomy, climate change mitigation and adaptation, vulnerability and resilience to climate change, natural resources management, and biodiversity conservation. He has consultancy experience across several countries in Africa, South America, the Caribbean, Asia, and North America. He has (co)authored over 130 publications (72 peer reviewed articles, over 30 conference papers and posters, and over 50 book chapters) in journals and books published by reputable publishers such as Elsevier, Springer, Taylor and Francis, MDPI, Wiley, and others. He equally serves as a reviewer and till date, he has reviewed over 650 manuscripts for journals of international repute such as *Communications Earth & Environment; Sustainable Production and Consumption; Environmental Impact Assessment Review; Geography and Sustainability; Sustainability Science; Regional Environmental Change*; *Plants*, *People*, *Planet; Agroforestry Systems*; *Climate and Development; International Journal of Agricultural Sustainability; Sustainability; Frontiers in Forests and Global Change; Forest Science and Technology; Frontiers in Plant Science; Frontiers in Genetics; Frontiers in Sustainable Food Systems*; *Forest and Society*. He sits on the Editorial Boards of Frontiers in Forests and Global Change, and Frontiers in Sustainable Food Production published by Frontiers; Discover Agriculture, Discover Conservation, and SN Social Sciences published by Springer Nature. He is a member of different national and international associations including the Cameroon Academy of Young Scientists—CAYS; British Ecological Society (BES) Grant Review Committee; International Union of Agroforestry (IUAF); International Union of Forest Research Organizations (IUFRO), YALI (Young African Leaders Initiative) Network; Congo Basin Young Agroforesters Association; Scholars Academic and Scientific Society (SAS) Young Research Fellow Member (SYRFM); International Society for Development and Sustainability (ISDS). From January 2023 till present, he is serving as Climate Change Consultant for FOKABS Canada. He is equally a Post-Doctoral Research Fellow with the EU funded TC4BE project coordinated by Wageningen University and Research—the Netherlands. He is affiliated to the following online research platforms:

- Academia (https://ubda.academia.edu/NyongPrincelyAwazi)
- Frontiers Loop (https://loop.frontiersin.org/people/1473281/overview)
- Google Scholar (https://scholar.google.com/citations?user=spxm5xQAAAAJ&hl=en)
- LinkedIn (https://www.linkedin.com/in/nyong-princely-awazi-552b87145/)
- ORCID (https://orcid.org/my-orcid?orcid=0000-0002-0801-0719)
- ResearchGate (https://www.researchgate.net/profile/Nyong-Princely-Awazi)
- Scopus (https://www.scopus.com/authid/detail.uri?authorId=57210996419)
- Web of Science (https://www.webofscience.com/wos/author/record/AAO-1853-2020)

List of Figures

List of Tables

CHAPTER 1

Introduction to building resilience to climate change for sustainable livelihoods in the Global South

Abstract Climate change poses a significant threat to the Global South, exacerbating vulnerabilities in regions already facing economic, social, and environmental challenges. The impacts of climate change such as rising temperatures, changing precipitation patterns, and increased frequency of extreme weather events disproportionately affect communities that rely heavily on natural resources for their livelihoods. These regions often face limited infrastructure, reduced adaptive capacity, and economic dependence on sectors such as agriculture, fisheries, forestry, and pastoralism, which are highly susceptible to climate variability. Vulnerability in these sectors is further compounded by poverty, inequality, and inadequate access to technological and financial resources. Resilience, in the context of climate change, refers to the ability of communities and ecosystems to absorb, adapt to, and recover from the shocks and stresses induced by a changing climate. Building resilience is essential to sustaining livelihoods, preserving biodiversity, and enhancing the capacity of communities to cope with future environmental changes. It involves strengthening adaptive capacities, diversifying economic activities, improving resource management, and fostering community-based solutions. The fisheries, forestry, pastoralism, and agriculture sectors are particularly critical in the Global South, as they form the backbone of many economies and provide essential services to local populations. Enhancing climate resilience in these sectors involves integrating climate-smart practices,

N. P. Awazi, *Building Resilience*, Palgrave Studies in Climate Resilient Societies, https://doi.org/10.1007/978-3-031-93791-0_1

promoting sustainable resource management, and ensuring that marginalized communities have access to the necessary tools, knowledge, and policies to navigate the challenges posed by climate change. Building resilience is therefore vital in sustaining these vital sectors and addressing climate vulnerabilities in the Global South.

Keywords Climate change · Global South · Vulnerability · Resilience · Livelihoods · Natural resources · Climate adaptation

1.1 Introduction to Climate Change in the Global South

Climate change is one of the most significant global challenges of the twenty-first century, and its impacts are disproportionately felt in the Global South (Hagen et al., 2022). The Global South refers to regions that are primarily situated in the Southern Hemisphere, encompassing Africa, Latin America, Asia, and small island developing states (Kowalski, 2021). These regions, despite contributing the least to the causes of climate change, are bearing the brunt of its consequences. The Global South is particularly vulnerable due to several factors, including economic dependence on natural resources, exposure to environmental stressors, limited adaptive capacity, and existing socio-economic disparities (Ngcamu, 2023). The phenomenon of climate change is not just an environmental issue; it is a deeply socio-economic challenge. Its impacts including rising temperatures, unpredictable weather patterns, sea-level rise, and increasing frequency of extreme weather events pose multifaceted threats to human livelihoods, ecosystems, and economic stability. However, the negative impacts of climate change are not felt equally across the globe. While wealthier nations, particularly those in the Global North, have the financial and technological resources to mitigate and adapt to these changes, the Global South remains highly vulnerable (Raj et al., 2022). A key reason for this vulnerability is the economic reliance on natural resources. Many countries in the Global South are heavily dependent on agriculture, fishing, and forestry—sectors that are directly affected by changing climate conditions. For example, Sub-Saharan Africa and parts of South Asia rely on rain-fed agriculture, making them extremely sensitive to changes in rainfall patterns and prolonged droughts (Omotoso et al., 2023; Tefera et al., 2025). The situation is compounded by the fact that climate change exacerbates existing environmental stressors such as deforestation, land degradation, and water scarcity.

The Global South's dependence on primary sectors like agriculture, mining, and fisheries exposes it to the direct impacts of climate change. Agriculture remains the backbone of many economies in the Global South, contributing significantly to GDP and providing livelihoods for the majority of the population (Jha, 2025). In Sub-Saharan Africa, for instance, over 60% of people depend on agriculture for their livelihood (Balgah et al., 2023; Hoffmann et al., 2022). Agriculture is vulnerable to changing climatic conditions, particularly irregular rainfall patterns, rising temperatures, and extreme weather events such as floods and droughts. Climate-induced changes in crop yields can lead to food insecurity, malnutrition, and increased poverty in rural areas, creating a feedback loop of economic stagnation and worsening vulnerability. Similarly, coastal and island nations, particularly in Southeast Asia and the Pacific, rely on fishing as a primary economic activity. Rising sea temperatures, ocean acidification, and the loss of coral reefs threaten marine biodiversity, with severe consequences for local economies and food security. According to the Intergovernmental Panel on Climate Change (IPCC), the global fisheries industry, a key component of many economies in the Global South, is expected to face a significant reduction in productivity due to climate change (IPCC, 2023). This could lead to a substantial decline in fish stocks, further exacerbating poverty in coastal communities.

The Global South is also disproportionately exposed to environmental stressors that amplify the effects of climate change. For example, many countries in the Global South are located in tropical and subtropical zones, where they are more prone to natural disasters such as cyclones, hurricanes, and floods. Low-lying nations like Bangladesh and the Maldives are particularly vulnerable to sea-level rise, which threatens to displace millions of people and submerge valuable land (Cardarelli & Pomper, 2024). Similarly, countries in Sub-Saharan Africa are highly susceptible to desertification and water scarcity, which threaten agriculture, livelihoods, and access to clean water. Moreover, environmental stressors often interact with pre-existing social, political, and economic vulnerabilities in the Global South, creating compounded risks. For instance, studies by Adom et al. (2022) and Leal Filho et al. (2022) found that climate change in Southern Africa has exacerbated existing water scarcity issues, leading to increased conflict over water resources, particularly between agricultural and industrial sectors. In addition, regions affected by armed conflict, political instability, or inadequate infrastructure are often less able to prepare for or respond to climate-related disasters. This lack of resilience further intensifies the social and economic impacts of climate change in these regions.

Adaptation to climate change is a complex process that requires significant resources, infrastructure, and institutional capacity. Unfortunately, many countries in the Global South face substantial challenges in this regard. These nations often have limited financial resources to invest in the technologies, infrastructures, and institutional frameworks required for effective climate change mitigation and adaptation. Without the necessary funding, technology, or expertise, countries in the Global South find themselves increasingly vulnerable to climate impacts. International climate finance mechanisms, such as the Green Climate Fund, have been established to assist developing nations in adapting to climate change. However, access to these funds has often been slow, and the mechanisms for distributing resources remain inadequate (Songwe et al., 2022). Furthermore, many countries in the Global South suffer from weak governance structures and corruption, which hinder the effective implementation of climate adaptation projects. These institutional weaknesses exacerbate the impacts of climate change, leaving these nations unable to mount an effective response to environmental challenges.

Climate change also disproportionately affects the most marginalized and vulnerable populations in the Global South, particularly women, children, indigenous communities, and the poor. These groups often have limited access to resources, education, and decision-making processes, which makes them more susceptible to the negative effects of climate change. For instance, rural women in Sub-Saharan Africa, who are responsible for collecting water and food for their families, face an increased burden as climate change worsens water scarcity and agricultural productivity. These social inequalities are compounded by the physical and environmental vulnerabilities that climate change exacerbates. The interconnections between poverty, inequality, and climate change are increasingly recognized as a central issue in the Global South. According to Birkmann et al. (2022), regions with lower adaptive capacities, high poverty rates, and poor access to healthcare and education are particularly vulnerable to the impacts of climate change. The most marginalized groups within these regions are often the ones who face the greatest risks, even though they contribute the least to greenhouse gas emissions.

Climate change is a global problem that requires global cooperation, and the Global South has long been at the forefront of pushing for climate justice. Developing nations have consistently argued that those most responsible for climate change, namely the industrialized countries in the Global North, should bear a greater responsibility for mitigating and adapting to its impacts. This has led to a focus on "climate justice" in international climate negotiations, with developing countries advocating for financial and technological support to address their unique vulnerabilities. The United Nations Framework Convention on Climate Change (UNFCCC) and the Paris Agreement have provided some hope for addressing the needs of the Global South, though there is considerable debate regarding the adequacy of these agreements in meeting the challenges of climate change. For instance, the commitment of developed countries to provide $100 billion annually for climate adaptation and mitigation in developing countries has faced criticism for being insufficient and difficult to implement effectively (Bos & Thwaites, 2021).

Climate change therefore represents a significant challenge for the Global South, where the impacts of a warming world are most keenly felt. These regions are particularly vulnerable due to their economic dependence on natural resources, exposure to environmental stressors, limited adaptive capacity, and existing socio-economic inequalities. The effects of climate change exacerbate pre-existing vulnerabilities, such as poverty, gender inequality, and political instability, creating a cycle of vulnerability that is difficult to break. While international mechanisms for addressing climate change exist, much more needs to be done to support the Global South in adapting to and mitigating the impacts of climate change. The challenge for policymakers, both within the Global South and the international community, is to ensure that climate change solutions are inclusive, just, and tailored to the unique needs and contexts of these vulnerable regions.

1.2 The Concept of Resilience in the Context of Climate Change in the Global South

The concept of resilience, when applied to climate change in the Global South, pertains to the ability of individuals, communities, and ecosystems to anticipate, adapt to, and recover from the impacts of climate stressors. These regions, often characterized by poverty, political instability, and social vulnerability, face disproportionate effects from climate change, including rising temperatures, changing precipitation patterns, extreme weather events, and sea-level rise. Resilience, in this context, is not simply about bouncing back from a shock; it encompasses adaptive capacities, anticipatory strategies, and the transformation of vulnerable systems into more robust, sustainable, and equitable ones. The importance of resilience in sustaining livelihoods and ecosystems in vulnerable communities in the Global South cannot be overstated, as it provides the foundation for building long-term adaptive capacities in the face of ongoing climate challenges.

Resilience can be defined as the ability of a system, whether ecological, social, or economic to absorb disturbance, adapt to change, and transform into a new state without losing its fundamental functions (Grygorenko & Naydonova, 2023). In the context of climate change, resilience entails not just recovering from climate-induced shocks, but also anticipating future changes and proactively making adjustments to avoid or mitigate the negative impacts. The term is often framed within a broader discourse of "adaptive capacity," which is the ability to anticipate, plan, and implement strategies that reduce the vulnerability of people, communities, and ecosystems to climate impacts (Pal, 2025). In the Global South, resilience is especially critical, as these regions are home to many of the world's poorest and most climate-vulnerable communities. For example, Africa, Latin America, and Asia are experiencing increased vulnerability to extreme weather events, such as droughts, floods, hurricanes, and heatwaves, which exacerbate existing socio-economic vulnerabilities. According to the Intergovernmental Panel on Climate Change (IPCC), regions in the Global South are projected to suffer the most severe climate impacts due to a combination of high exposure to climate hazards, low adaptive capacity, and already-existing challenges related to poverty, weak governance, and limited access to resources (IPCC, 2023). Resilience in these contexts is not simply about maintaining the status quo, but also involves transformative changes in how societies organize, govern, and interact with the environment. It calls for enhancing the capacity of vulnerable populations to absorb climate shocks, adapt their livelihoods, and transform their societies to cope with future challenges.

Resilience emphasizes a preferred development trajectory, typically assessed through indicators like yield, agricultural output, household income, or GDP per capita. This trajectory is then impacted by a stressor or shock. A stressor might stem from rising maximum or minimum temperatures, the impacts of a progressively drier climate, or increasing sea levels while a shock refers to an unexpected and abrupt event, such as a significant flood, drought, storm surge, or cyclone (Fig. 1.1). Countermeasures that enhance resilience can vary widely, encompassing social, economic, and technological approaches, such as institutional (land-use zoning, integrated warning systems), economic (weather insurance, microcredit programs), physical (cyclone shelters, embankments), medical (vaccines), environmental (mangrove shelterbelts), agricultural (drought- and flood-resistant crop varieties, enhanced nutrient use), livelihood (income diversification, rural–urban connections, mobility for migration), and education (dissemination of information and skill development).

In the Global South, many livelihoods are heavily dependent on agriculture, natural resources, and informal economies that are particularly vulnerable to climate change. Small-scale farmers, fisherfolk, pastoralists, and indigenous populations are often on the frontlines of climate change, experiencing disruptions to food production, water supply, and land use. For instance, droughts can reduce crop yields, while floods and cyclones can destroy homes and infrastructure. Such disruptions not only threaten immediate survival but also undermine long-term development prospects. Building resilience within these communities is vital for sustaining their livelihoods. Adaptive capacity involves the ability of individuals and communities to make adjustments to their livelihoods

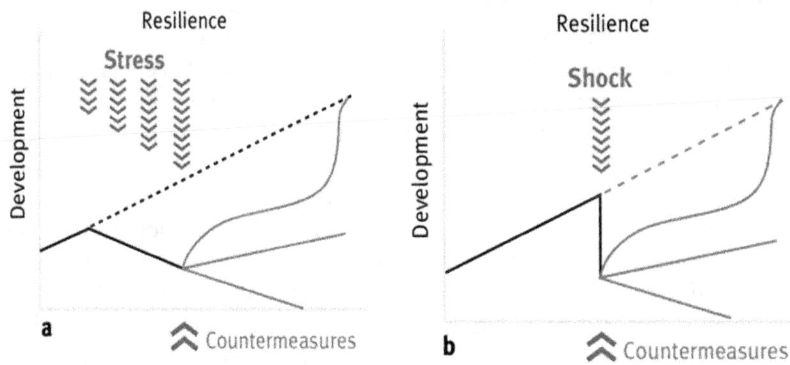

Fig. 1.1 Patterns of resilience (a) to stress, (b) to shock (*Source* Conway, 2009)

in response to changing climatic conditions. This might include diversifying crops, adopting climate-smart agricultural techniques, improving water management practices, and enhancing early warning systems for extreme weather events. In Sub-Saharan Africa, for example, farmers are increasingly adopting drought-resistant crops and using conservation agriculture techniques to improve soil health and water retention (Okoronkwo et al., 2024). Moreover, resilience is crucial for protecting the most vulnerable groups in these communities, such as women, children, and the elderly, who are disproportionately impacted by climate stressors. Women, in particular, are often responsible for household food production, water collection, and care work, and their roles are exacerbated during times of climate-induced stress. Strengthening women's resilience through capacity-building initiatives, access to resources, and decision-making power can significantly improve community resilience overall (Chikwe et al., 2024). In addition, resilience is closely linked to the broader concept of sustainable development. While short-term coping strategies may help individuals weather immediate climate impacts, long-term resilience requires structural changes that promote equitable development. This includes improving infrastructure, enhancing education and health services, and creating economic opportunities that reduce dependence on climate-sensitive sectors. Sustainable livelihoods, therefore, are a key element of resilience and require multi-dimensional approaches that address social, economic, and environmental factors simultaneously (Quandt & Paderes, 2023).

Beyond human systems, resilience in the context of climate change also extends to ecosystems, particularly in biodiversity-rich areas in the Global South, such as tropical rainforests, mangroves, and coral reefs (Fig. 1.2). These ecosystems provide essential services, including carbon sequestration, water filtration, and habitat for biodiversity. However, they are under increasing threat from climate change, deforestation, land degradation, and overexploitation. Resilience in ecosystems is essential not only for environmental sustainability but also for the continued well-being of communities dependent on ecosystem services. For example, many coastal communities in the Global South rely on mangroves for protection against storm surges and coastal erosion. Mangroves also provide important resources for fishing and agriculture. Yet, rising sea levels and deforestation are depleting these ecosystems, leaving communities more vulnerable to climate hazards (Bhowmik et al., 2022). To enhance ecosystem resilience, it is necessary to integrate ecosystem-based adaptation (EbA) strategies that restore and protect ecosystems while building adaptive capacity. This approach includes measures such as reforestation, sustainable land management, and the restoration of degraded ecosystems to improve both biodiversity and the resilience of human communities. The restoration of wetlands in coastal areas, for instance, can help buffer communities from rising sea levels and reduce the risk of flooding, while also providing livelihood opportunities through ecotourism and sustainable harvesting of resources (Guerry et al., 2022; Hernández-Delgado, 2024). Furthermore, the preservation and restoration of ecosystems in the Global South are not only crucial for local adaptation but also have global implications for climate change mitigation. Many ecosystems in the Global South, such as the Amazon rainforest and the Congo Basin, are important carbon sinks. Protecting these ecosystems from degradation is therefore integral to addressing the global challenge of climate change.

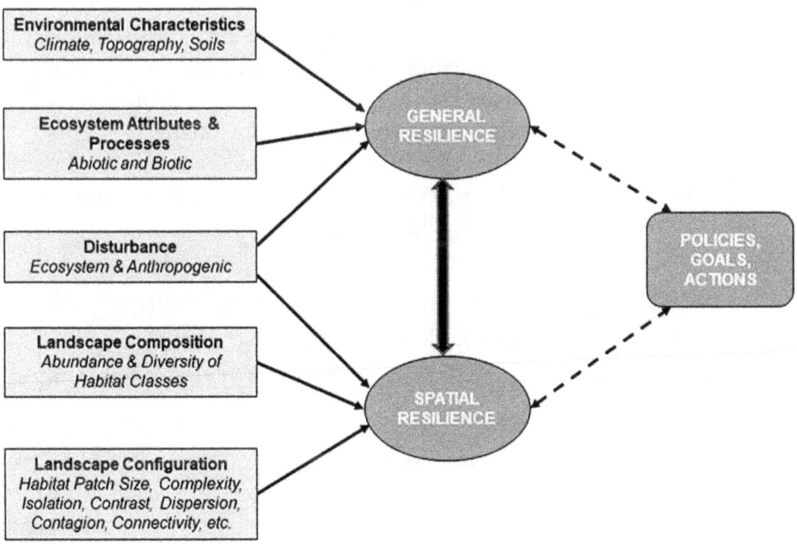

Fig. 1.2 Ecological resilience concepts for the Global South (*Source* Chambers et al., 2019)

The effectiveness of resilience-building strategies in the Global South depends heavily on governance structures and social capital. Effective governance involves the establishment of policies, institutions, and processes that enable communities to respond to climate change. This includes strengthening local governance, enhancing the capacity of governments to manage climate risks, and promoting policies that prioritize climate adaptation and resilience. Social capital, which refers to the networks, norms, and trust that enable people to work together, plays a critical role in enhancing resilience. Communities with strong social ties are better able to share resources, knowledge, and support during times of crisis. For example, indigenous knowledge systems, which are often rooted in a deep understanding of local ecosystems and climate patterns, can be critical for developing locally appropriate solutions to climate challenges. Recognizing and integrating this knowledge into climate policies and adaptation strategies is essential for building resilience in the Global South (Ziervogel et al., 2022). International support is also critical in building resilience. While the Global South faces significant challenges in addressing climate change, financial and technical assistance from developed countries and international organizations can play a key role in enhancing adaptive capacities. The Green Climate Fund and other international mechanisms are designed to provide financial support to vulnerable countries, but the effectiveness of these mechanisms depends on ensuring that the funds are accessible, transparent, and aligned with the needs of local communities (Pill, 2022).

1.3 Overview of Climate Resilience in the Fisheries, Forestry, Pastoralism, and Agriculture Sectors in the Global South

Climate change is one of the most pressing global challenges of our time, disproportionately affecting vulnerable communities in the Global South. These regions, which include countries in Sub-Saharan Africa, Latin America, Asia, and Oceania, often depend on natural resource-based livelihoods such as fisheries, forestry, pastoralism, and agriculture. These sectors are highly sensitive to shifts in climate variables, including temperature, rainfall, and extreme weather events which necessitates different resilience mechanisms to buffer the impacts of climate change (Table 1.1). Climate resilience, understood as the capacity to anticipate, prepare for, respond to, and recover from climate-related shocks, has thus become a critical concept in shaping sustainable development strategies in these regions.

Fisheries in the Global South are central to the livelihoods of millions of people, providing food security and income, especially in coastal and inland areas. However, these sectors are highly vulnerable to climate change. Rising ocean temperatures, ocean acidification, and changing precipitation patterns all impact fish stocks and aquatic ecosystems, disrupting local economies and food security (Iyiola et al., 2022). For instance, the increase in sea temperatures has led to the migration of fish species, often to cooler waters, leaving traditional fishing communities without their primary sources of income (Muringai et al., 2022). Coral reefs, which support numerous fish species, are increasingly threatened by bleaching events due to warmer waters (Obura et al., 2022). In the Global South, the adaptive capacity of fisheries communities is often limited by poor infrastructure, low levels of education, and inadequate access to technology. In response, resilience-building strategies have been explored through various mechanisms, including improved resource management, ecosystem restoration, and diversifying livelihoods. One example is the promotion of community-based fisheries management (CBFM) models, where local communities are empowered to manage their fisheries resources. CBFM has shown success in several regions, such as in the Philippines, where localized management practices have helped restore fish stocks and improve the livelihoods of fishers (Marriott et al., 2021).

The forestry sector in the Global South plays an essential role in sustaining biodiversity, mitigating climate change through carbon sequestration, and providing resources for local communities. Forests also contribute to the livelihoods of rural populations through timber, fuelwood, and non-timber forest products. However, deforestation and forest degradation, exacerbated by climate change, threaten the sustainability of these forests. Rising temperatures and changing precipitation patterns can reduce forest growth and biodiversity, while increasing the frequency of forest fires and pests (Robinne & Secretariat, 2021). For instance, the Amazon rainforest, which is crucial for global climate regulation, has experienced increasing droughts and forest fires, driven in part by climate change and land-use practices (Celis et al., 2023). To enhance climate resilience in the forestry sector, several adaptation strategies have been implemented. These include the restoration of degraded lands through

Table 1.1 Climate change impacts and resilience strategies in different livelihood sectors in the Global South

Sector	*Climate change impacts*	*Resilience strategies*	*Key references*
Fisheries	Vulnerable to rising ocean temperatures, ocean acidification, changing precipitation, and fish species migration. Coral reefs are threatened by bleaching, disrupting local economies and food security	Community-based fisheries management (CBFM), resource management improvements, ecosystem restoration, livelihood diversification	Iyiola et al. (2022), Muringai et al. (2022), Obura et al. (2022), Marriott et al. (2021)
Forestry	Deforestation, forest degradation, rising temperatures, changing precipitation, forest fires, and pests threaten biodiversity and livelihoods	Agroforestry, reforestation, afforestation, sustainable forest management, community forestry, local governance enhancement	Robinne and Secretariat (2021), Celis et al. (2023), Castle et al. (2021), Ghimire et al. (2024), Murer and Piccoli (2022)
Pastoralism	Changing rainfall patterns, increasing temperatures, droughts, desertification, and grazing land degradation. Mobility is restricted due to land-use changes, conflicts, and water scarcity	Management of mobility through grazing corridors, water-sharing agreements, drought-resistant livestock, early warning systems, livelihood diversification (crop farming, livestock product markets), access to climate information	Bogale and Erena (2022), Sulieman and Young (2023), Amburgey (2024), Mbaziira et al. (2023), Abdela (2024), Kassegn and Endris (2021), Habib et al., (2023), Boas (2022)
Agriculture	Vulnerable to shifting rainfall, droughts, erratic weather events, and impacts on crop yields and food security. Smallholder farmers face limited access to resources, technology, and markets	Climate-smart agriculture (CSA), drought-tolerant crops, improved irrigation, conservation tillage, crop rotation, intercropping, strengthening institutional frameworks, social safety nets (insurance, cash transfers)	Bedeke (2023), Mensah et al. (2021), Adimassu et al. (2024), Ariom et al. (2022), Fitrinitia and Matsuyuki (2023)

reforestation and afforestation, the promotion of agroforestry systems, and the adoption of sustainable forest management practices. Agroforestry, which integrates trees into agricultural landscapes, offers a means of reducing the vulnerability of farming communities to climate impacts while providing additional income streams. In the Global South, where subsistence agriculture is widespread, agroforestry practices can diversify income and improve food security while protecting soils and water resources (Castle et al., 2021; Ghimire et al., 2024). In addition to these strategies, forest governance plays a critical role in fostering resilience. Local communities are often the stewards of forests, and ensuring their participation in decision-making processes is crucial for long-term sustainability. The concept of community forestry, in which local populations actively manage forest resources, has been increasingly promoted in countries like Nepal, India, and Kenya (Murer & Piccoli, 2022). By enhancing local governance structures, these initiatives aim to build both ecological and social resilience.

Pastoralism, or the practice of raising livestock, is a key livelihood strategy in arid and semi-arid regions of the Global South, particularly in Sub-Saharan Africa, the Sahel, and parts of Asia. However, climate change presents significant challenges to pastoral systems, as changing rainfall patterns and increasing temperatures lead to droughts, desertification, and the degradation of grazing lands (Bogale & Erena, 2022). The mobility of pastoralists, which is a key strategy for adapting to environmental variability, is increasingly constrained by land-use changes, conflicts, and the growing scarcity of water and forage (Sulieman & Young, 2023). Pastoralist resilience to climate change is enhanced through various strategies. One important approach is the management of pastoral mobility through community-driven approaches, which include the establishment of grazing corridors, water-sharing agreements, and collective resource management practices (Amburgey, 2024). In East Africa, the implementation of mobile pastoralism as a form of adaptive management has allowed communities to maintain livestock productivity despite variable climatic conditions (Mbaziira et al., 2023). Moreover, pastoralists are increasingly integrating drought-resistant livestock breeds, improved livestock management practices, and early warning systems to reduce the vulnerability of their herds (Abdela, 2024). Another significant strategy is the diversification of livelihoods, such as the inclusion of crop farming or the development of markets for livestock products (Habib et al., 2023;

Kassegn & Endris, 2021). Diversifying income sources allows pastoralists to buffer the economic shocks caused by climatic extremes, such as drought. Additionally, strengthening the access of pastoralists to climate information through technology and local networks has been shown to improve their decision-making and coping strategies (Boas, 2022).

Agriculture is a cornerstone of the economies of many countries in the Global South, particularly in Sub-Saharan Africa and South Asia. This sector is highly vulnerable to the impacts of climate change, with shifting rainfall patterns, prolonged droughts, and erratic weather events threatening crop yields and food security. Smallholder farmers, who make up a significant proportion of the agricultural workforce in these regions, are particularly vulnerable due to limited access to resources, technology, and markets (Bedeke, 2023). Building climate resilience in agriculture involves a combination of strategies aimed at improving productivity, sustainability, and risk management. Climate-smart agriculture (CSA) has emerged as a key framework for enhancing resilience, which focuses on three pillars: increasing agricultural productivity and income, adapting to climate change, and reducing greenhouse gas emissions (Adimassu et al., 2024; Mensah et al., 2021). CSA strategies include the use of drought-tolerant crop varieties, improved irrigation techniques, conservation tillage, and agroecological practices such as crop rotation and intercropping (Ariom et al., 2022). In addition to these technological innovations, there is growing recognition of the importance of strengthening the institutional and policy frameworks that support agriculture. Policies that facilitate access to climate information, markets, and financial services are essential for enabling farmers to adapt to climate change. Furthermore, social safety nets, such as crop insurance schemes and cash transfers, have been increasingly implemented to support vulnerable farmers during climate shocks (Fitrinitia & Matsuyuki, 2023). These mechanisms help reduce the risks associated with climate variability and provide a buffer for farmers who are affected by extreme weather events.

1.4 Key Themes

While climate change is a global phenomenon, its impacts are felt most acutely by vulnerable populations in developing nations. These communities often face heightened risks due to their geographic location, dependence on climate-sensitive sectors such as agriculture, and limited capacity to adapt to environmental stresses. Therefore, building resilience in the Global South is essential for ensuring sustainable livelihoods and long-term development. Key themes such as adaptation, sustainable livelihoods, policy frameworks, local solutions, and the Global South are examined.

1.4.1 *Climate Change Adaptation in the Global South*

Adaptation refers to the adjustments or changes in practices, processes, and structures to minimize the negative impacts of climate change or take advantage of any potential opportunities. In the Global South, adaptation strategies are crucial for managing the risks of climate change, especially given the region's limited financial and technical resources. A key theme for this book is the examination of bottom-up adaptation approaches. These approaches empower local communities to identify their vulnerabilities and develop context-specific strategies to cope with climate-related challenges. Studies such as those of Shammin et al. (2022) and Choudhury et al. (2021) emphasize the importance of local knowledge and community-based adaptation in improving resilience. This is particularly relevant in rural areas where traditional agricultural practices are deeply tied to local knowledge systems, which can offer sustainable solutions to climate challenges. For example, agroecological practices have shown resilience to climate stressors by integrating biodiversity and soil health with farming practices. Additionally, the integration of adaptation with development objectives is a crucial theme. Climate change adaptation must not be viewed in isolation but as part of broader development goals. This means adapting development strategies to be climate-resilient. Friedman (2023) and Muccione et al. (2024) argue that successful adaptation hinges on transforming development pathways to address both climate risks and the social and economic inequalities that make communities more vulnerable. Another critical area of adaptation is ecosystem-based adaptation (EbA), which uses biodiversity and ecosystem services as part of an overall strategy to help communities adapt to the adverse effects of climate change. EbA has proven effective in the Global South, particularly in coastal and forested areas, where the protection of mangroves or reforestation can enhance both community resilience and biodiversity.

1.4.2 *Sustainable Livelihoods in the Global South*

The concept of sustainable livelihoods refers to the ability of individuals or communities to access resources, manage them wisely, and ensure their long-term well-being in the face of changing environmental conditions (Natarajan et al., 2022). In the Global South, livelihoods are often heavily dependent on natural resources, such as agriculture, fisheries, and forestry. These sectors are highly vulnerable to climate change, making the need for sustainable livelihood strategies even more pressing. Livelihoods must be diverse, flexible, and adaptive to changing climatic conditions. Climate change challenges the sustainability of traditional livelihoods, and the book will examine how communities in the Global South can build diversified, climate-resilient livelihoods through alternative income-generating activities, such as ecotourism, agroforestry, and renewable energy. Livelihood diversification is a key strategy for reducing vulnerability to climate risks. Diversification allows communities to spread risk across various income sources, making them less susceptible to the devastating effects of climate change on any single sector. As Mohammed et al. (2021) and Habib et al., (2023) suggests, diversification is an essential component of rural resilience, particularly in agricultural communities in the Global South. Through the incorporation of non-farm activities such as small-scale manufacturing or services, communities are better positioned to withstand climate shocks. Another critical element is land tenure and resource access. In many parts of the Global South, land tenure is insecure, especially among marginalized groups, such as women or indigenous populations. This insecurity limits access to resources, making it harder for communities to adapt to climate impacts. Santpoort et al. (2021) and Shahin et al. (2022) argue that empowering women and marginalized groups through secure land tenure can increase resilience, as it allows them to make long-term investments in land and other resources. Thus, land rights and access to natural resources must be central to any resilience-building strategies.

1.4.3 *Policy Frameworks for Climate Change Resilience in the Global South*

Effective policy is vital in building resilience to climate change, particularly in the Global South, where governments may struggle to provide adequate resources and infrastructure. The book will explore the role of policy frameworks in supporting adaptation and resilience-building at local, national, and international levels. A crucial theme is the mainstreaming of climate change into development policy. As argued by Hurlimann et al. (2021) and Mani and Goniewicz (2023), climate change must be incorporated into national development plans and policies to ensure that adaptation and mitigation measures are not viewed as isolated activities. Integrating climate change considerations into national planning processes helps ensure that investments and strategies are climate-resilient, enhancing overall development outcomes. The book addresses international climate finance, which is critical for the Global South. The Green Climate Fund (GCF) and other international financial mechanisms have emerged as key sources of funding for climate adaptation projects. However, Kalaidjian and Robinson (2022) and Skovgaard et al. (2023) argue that the effectiveness of international climate finance is hindered by bureaucratic complexities and the limited ability of local governments and communities to access these funds. Hence, an important theme will be examining how policy can enable better access to climate finance at the local level, ensuring that funds reach the most vulnerable populations. Disaster risk reduction (DRR) is another essential policy area, as it focuses on reducing the risks posed by natural disasters, which are exacerbated by climate change. The Sendai Framework for Disaster Risk Reduction emphasizes the importance of building resilience through proactive measures, such as early warning systems and disaster preparedness plans. The book will explore how national governments in the Global South can strengthen DRR policies and implement them effectively, incorporating community participation in decision-making.

1.4.4 *Local Solutions in the Global South*

One of the most powerful ways to build resilience in the Global South is through local solutions. These solutions arise from the needs, knowledge, and capacities of communities themselves. Studies by Karki et al. (2021) and Patnaik (2021) emphasize the importance of community-based adaptation (CBA), which recognizes that local communities are best positioned to identify and address their unique vulnerabilities. A theme explored in the book will be local innovation and knowledge-sharing. Communities in the Global South have long been experts in managing environmental risks through traditional ecological knowledge (TEK). For instance, indigenous peoples have developed sophisticated land and water management practices, which are often more sustainable than industrial alternatives. The book highlights case studies where local knowledge is combined with scientific approaches to create effective climate adaptation strategies, such as the revival of traditional water harvesting techniques or the use of native crop varieties that are more resilient to changing weather patterns. Furthermore, participatory decision-making will be emphasized as a key component of successful local adaptation. Eicken et al. (2021) argue that top-down approaches to climate change adaptation often fail to consider the needs and priorities of local communities. Therefore, ensuring that communities have a voice in shaping policies and solutions is essential to building resilience.

1.4.5 Global South

The Global South refers to countries in the southern hemisphere, primarily in Africa, Latin America, Asia, and parts of Oceania, which have historically been characterized by lower levels of economic development and political power compared to the Global North (Pinheiro, 2024). This region is disproportionately affected by the impacts of climate change, economic inequality, and historical colonial exploitation. Studies by Owolabi (2023) and Vertigans (2021) have highlighted the structural inequalities that persist between the Global South and North, emphasizing how colonial legacies continue to shape development challenges in the South. The Global South is also home to significant resilience and innovation, with many countries adopting unique strategies to address climate change, economic development, and social justice. Galappaththi et al. (2021) discusses how local knowledge systems and community-based adaptation strategies in the Global South can offer valuable insights for tackling global issues like climate change. Despite the challenges, the Global South has demonstrated great potential for transformation through South-South cooperation (Banik & Mawdsley, 2023), where countries in the region share resources, knowledge, and solutions to common problems. The Global South remains a crucial part of the global discourse on sustainable development, equity, and climate justice. A significant theme in the book is the exploration of unique challenges and opportunities of the Global South in addressing climate change. While the region faces immense challenges, such as limited resources, political instability, and historical inequalities, it also holds opportunities for innovative and transformative solutions. The book also explores how South-South cooperation can play a role in climate resilience. Many countries in the Global South have developed valuable knowledge and practices that can be shared across borders. For example, countries like Brazil, India, and Kenya have implemented successful climate adaptation programs that can serve as models for others in similar contexts.

1.5 Structure of the Book

The book "Building Resilience: Climate Change and Livelihoods in the Global South" provides a comprehensive exploration of the challenges and opportunities related to climate resilience across various sectors in the developing world. This chapter sets the stage by examining climate change and its disproportionate impacts on the Global South. It introduces the concept of climate resilience, emphasizing its critical importance for livelihoods in vulnerable communities. This chapter also provides an overview of resilience in key sectors such as fisheries, forestry, pastoralism, and agriculture, offering a preview of the sectors explored in greater depth throughout the book. Central to the discussion are themes of adaptation, sustainable livelihoods, policy development, and local solutions, all framed within the context of the Global South. In Chapter 2, the focus shifts to forest-dependent communities, delving into the relationship between these communities and forest ecosystems. It highlights the climate impacts on forests, such as deforestation and degradation, and explores adaptive strategies employed by local populations to sustain their livelihoods. Through case studies, the chapter also examines community-led conservation approaches and offers policy implications for enhancing forest sustainability and resilience. Chapter 3 addresses climate resilience in fishing communities, focusing on the significant challenges posed by climate change to marine ecosystems. Issues such as ocean acidification, sea-level rise, and shifting fish populations are explored, with an emphasis on community-based fisheries management and livelihood diversification as adaptive strategies. The chapter identifies gaps in policy frameworks and highlights the need for better governance and enhanced community involvement in fisheries management. Chapter 4 examines pastoralist livelihoods, emphasizing their particular vulnerability to climate change. It discusses how pastoralists are impacted by changing weather patterns, loss of grazing land, and water scarcity. The chapter also highlights traditional resilience practices, technological innovations, and sustainability prospects, while assessing resilience frameworks and policies designed to support pastoralist communities. Chapter 5 shifts attention to agricultural communities in the Global South, exploring how recurrent climate events such as droughts and floods affect agricultural productivity and food security. It investigates adaptive agricultural practices like crop diversification, conservation agriculture, and the use of climate-resilient crops. The chapter also considers the role of education, access to resources, and technology in enhancing resilience within agricultural systems. In Chapter 6, the book reviews climate resilience policies across the fisheries, agriculture, and forestry sectors, evaluating their effectiveness and the challenges in their implementation. It introduces emerging trends like ecosystem-based adaptation and climate-smart agriculture and discusses how these approaches could shape future policy. The chapter concludes with recommendations for integrated, inclusive, and locally informed policy frameworks to enhance resilience in these sectors. The final chapter, Chapter 7, synthesizes the key findings from the previous chapters, reflecting on resilience-building strategies and the interconnectedness of the sectors discussed. It underscores the importance of integrated approaches and the role of both international cooperation and local action in enhancing resilience. The chapter concludes with a call for global commitments, inclusive policies, and sustainable practices to strengthen climate resilience in the Global South, as well as suggestions for future research and actionable steps. Together, the seven (07) chapters provide a holistic and multi-sectoral approach to understanding and addressing the impacts of climate change on livelihoods in the Global South, offering insights, solutions, and policy recommendations for building long-term resilience in the face of a changing climate.

References

Abdela, U. (2024). Assessment of community-driven drought risk management strategies in pastoral and agro-pastoral district of Bale zones south east Ethiopia. *Frontiers in Environmental Science, 12*, 1411138.

Adimassu, Z., Tamene, L., Tibebe, D., Ebrahim, M., & Abera, W. (2024). Identification and prioritization of context-specific climate smart agricultural (CSA) practices in Ethiopia: A framework approach. *Environmental Research Communications, 6*(6), Article 065007.

Adom, R. K., Simatele, M. D., & Reid, M. (2022). The threats of climate change on water and food security in South Africa. *American Journal of Environment and Climate, 1*(2), 6–10.

Amburgey, E. (2024). *Migration and translocal livelihoods: Transformation amidst climate change in Mustang, Nepal* (Doctoral dissertation, University of British Columbia).

Ariom, T. O., Dimon, E., Nambeye, E., Diouf, N. S., Adelusi, O. O., & Boudalia, S. (2022). Climate-smart agriculture in African countries: A Review of strategies and impacts on smallholder farmers. *Sustainability, 14*(18), 11370.

Balgah, R. A., Ngwa, K. A., Buchenrieder, G. R., & Kimengsi, J. N. (2023). Impacts of floods on agriculture-dependent livelihoods in Sub-saharan Africa: An assessment from multiple geo-ecological zones. *Land, 12*(2), 334.

Banik, D., & Mawdsley, E. (2023). South-South Cooperation and global development in a multipolar world: China and India in Africa. *Journal of International Development*. https://doi.org/10.1002/jid.3789

Bedeke, S. B. (2023). Climate change vulnerability and adaptation of crop producers in sub-Saharan Africa: A review on concepts, approaches and methods. *Environment, Development and Sustainability, 25*(2), 1017–1051.

Bhowmik, A. K., Padmanaban, R., Cabral, P., & Romeiras, M. M. (2022). Global mangrove deforestation and its interacting social-ecological drivers: A systematic review and synthesis. *Sustainability, 14*(8), 4433.

Birkmann, J., Liwenga, E., Pandey, R., Boyd, E., Djalante, R., Gemenne, F., Leal Filho, W., Pinho, P. F., Stringer, L., & Wrathall, D. (2022). Poverty, livelihoods and sustainable development. In H.-O. Pörtner, D. C. Roberts, M. Tignor, E. S. Poloczanska, K. Mintenbeck, A. Alegría, M. Craig, S. Langsdorf, S. Löschke, V. Möller, A. Okem, & B. Rama (Eds.), *Climate Change 2022: Impacts, adaptation and vulnerability.* Contribution of Working Group II to the Sixth Assessment Report of the Intergovernmental Panel on Climate Change (pp. 1171–1274). Cambridge University Press. https://doi.org/10.1017/9781009325844.010

Boas, I. (2022). The diversification of pastoralist herding: Navigating socio-climatic risk via mobile technologies. *Journal of Ethnic and Migration Studies, 48*(14), 3433–3449.

Bogale, G. A., & Erena, Z. B. (2022). Drought vulnerability and impacts of climate change on livestock production and productivity in different agro-Ecological zones of Ethiopia. *Journal of Applied Animal Research, 50*(1), 471–489.

Bos, J., & Thwaites, J. (2021). A breakdown of developed countries' public climate finance contributions towards the $100 billion goal. *World Resources Institute*, 2021-10.

Cardarelli, R., & Pomper, H. (2024). Climate-driven migration–Asia (Case studies of the Maldives and Bangladesh). In *Children and the climate migration crisis: A casebook for global climate action in practice and policy* (pp. 99–118). Emerald Publishing Limited.

Castle, S. E., Miller, D. C., Ordonez, P. J., Baylis, K., & Hughes, K. (2021). The impacts of agroforestry interventions on agricultural productivity, ecosystem services, and human well-being in low-and middle-income countries: A systematic review. *Campbell Systematic Reviews, 17*(2), Article e1167.

Celis, N., Casallas, A., Lopez-Barrera, E. A., Felician, M., De Marchi, M., & Pappalardo, S. E. (2023). Climate change, forest fires, and territorial dynamics in the Amazon rainforest: An integrated analysis for mitigation strategies. *ISPRS International Journal of Geo-Information, 12*(10), 436.

Chambers, J. C., Allen, C. R., & Cushman, S. A. (2019). Operationalizing ecological resilience concepts for managing species and ecosystems at risk. *Frontiers in Ecology and Evolution, 7*, 241.

Chikwe, C. F., Kuteesa, C. F., & Ediae, A. A. (2024). Gender equality advocacy and socio-economic inclusion: A comparative study of community-based approaches in promoting women's empowerment and economic resilience (2022). *International Journal of Scientific Research Updates, 8*(2), 110–121.

Choudhury, M. U. I., Haque, C. E., Nishat, A., & Byrne, S. (2021). Social learning for building community resilience to cyclones: Role of indigenous and local knowledge, power, and institutions in coastal Bangladesh. *Ecology & Society, 26*(1).

Conway, G. (2009). *The science of climate change in Africa: Impacts and adaptation. Grantham Institute for Climate Change* (Discussion Paper No. 1), Imperial College London. https://www.mamopanel.org/media/uploads/files/SCIENCE_OF_CLIMATE_CHANGE_IN_AFRICA-_IMPACTS_AND_ADAPTION_2009.pdf

Eicken, H., Danielsen, F., Sam, J. M., Fidel, M., Johnson, N., Poulsen, M. K., & Enghoff, M. (2021). Connecting top-down and bottom-up approaches in environmental observing. *BioScience, 71*(5), 467–483.

Fitrinitia, I. S., & Matsuyuki, M. (2023). Social protection for climate-disasters: A case study of the program Keluarga Harapan cash transfer program for smallholder farm household in Indonesia. *Progress in Disaster Science, 17*, Article 100278.

Friedman, E. (2023). Constructing the adaptation economy: Climate resilient development and the economization of vulnerability. *Global Environmental Change, 80*, Article 102673.

Galappaththi, E. K., Ford, J. D., Bennett, E. M., & Berkes, F. (2021). Adapting to climate change in small-scale fisheries: Insights from indigenous communities in the global north and south. *Environmental Science & Policy, 116*, 160–170.

Ghimire, M., Khanal, A., Bhatt, D., Dahal, D., & Giri, S. (2024). Agroforestry systems in Nepal: Enhancing food security and rural livelihoods—A comprehensive review. *Food and Energy Security, 13*(1), Article e524.

Grygorenko, Z., & Naydonova, G. (2023). The concept of "resilience": History of formation and approaches to definition. *Public Administration and Law Review, 2*, 76–88.

Guerry, A. D., Silver, J., Beagle, J., Wyatt, K., Arkema, K., Lowe, J., & Sharma, J. (2022). Protection and restoration of coastal habitats yield multiple benefits for urban residents as sea levels rise. *npj Urban Sustainability, 2*(1), 13.

Habib, N., Ariyawardana, A., & Aziz, A. A. (2023). The influence and impact of livelihood capitals on livelihood diversification strategies in developing countries: A systematic literature review. *Environmental Science and Pollution Research, 30*(27), 69882–69898.

Hagen, I., Huggel, C., Ramajo, L., Chacón, N., Ometto, J. P., Postigo, J. C., & Castellanos, E. J. (2022). Climate change-related risks and adaptation potential in Central and South America during the 21st century. *Environmental Research Letters, 17*(3), Article 033002.

Hernández-Delgado, E. A. (2024). Coastal restoration challenges and strategies for small island developing states in the face of sea level rise and climate change. *Coasts, 4*(2), 235–286.

Hoffmann, R., Wiederkehr, C., Dimitrova, A., & Hermans, K. (2022). Agricultural livelihoods, adaptation, and environmental migration in sub-Saharan drylands: A meta-analytical review. *Environmental Research Letters, 17*(8), Article 083003.

Hurlimann, A., Moosavi, S., & Browne, G. R. (2021). Urban planning policy must do more to integrate climate change adaptation and mitigation actions. *Land Use Policy, 101*, Article 105188.

IPCC. (2023). Summary for policymakers. In Core Writing Team, H. Lee, & J. Romero (Eds.), *Climate Change 2023: Synthesis Report. Contribution of Working Groups I, II and III to the Sixth Assessment Report of the Intergovernmental Panel on Climate Change* (pp. 1–34). IPCC. https://doi.org/10.59327/IPCC/AR6-9789291691647.001

Iyiola, A. O., Asiedu, B., Oyewole, E. O., & Akinrinade, A. J. (2022). Impacts of climate change on aquatic biodiversity in Africa. In *Biodiversity in Africa: Potentials, threats and conservation* (pp. 369–394). Springer Nature Singapore.

Jha, P. (2025). Challenges of decent work for smallholders and agricultural workers: Reflections on the Global South. In *The Elgar companion to decent work and the Sustainable Development Goals* (pp. 247–258). Edward Elgar Publishing.

Kalaidjian, E., & Robinson, S. A. (2022). Reviewing the nature and pitfalls of multilateral adaptation finance for small island developing states. *Climate Risk Management, 36*, Article 100432.

Karki, G., Bhatta, B., Devkota, N. R., Acharya, R. P., & Kunwar, R. M. (2021). Climate change adaptation (CCA) interventions and indicators in Nepal: Implications for sustainable adaptation. *Sustainability, 13*(23), 13195.

Kassegn, A., & Endris, E. (2021). Review on livelihood diversification and food security situations in Ethiopia. *Cogent Food & Agriculture, 7*(1), 1882135.

Kowalski, A. M. (2021). Global south-global north differences. *No poverty* (pp. 389–400). Springer International Publishing.

Leal Filho, W., Totin, E., Franke, J. A., Andrew, S. M., Abubakar, I. R., Azadi, H., & Global Adaptation Mapping Initiative Team. (2022). Understanding responses to climate-related water scarcity in Africa. *Science of the Total Environment, 806*, 150420.

Mani, Z. A., & Goniewicz, K. (2023). Adapting disaster preparedness strategies to changing climate patterns in Saudi Arabia: A rapid review. *Sustainability, 15*(19), 14279.

Marriott, S. E., Cox, C., Amolo, R. C., Apistar, D., Mancao, R. H., & de Mutsert, K. (2021). Implications of community-based management of marine reserves in the Philippines for reef fish communities and biodiversity. *Frontiers in Marine Science, 8*, Article 731675.

Mbaziira, J., Egeru, A., Bamutaze, Y., Kisira, Y., & Nabiroko, M. (2023). Assessing the dynamics of agropastoral farmers' adaptive capacity to drought in Uganda's cattle corridor. *Climate Risk Management, 41*, Article 100545.

Mensah, H., Ahadzie, D. K., Takyi, S. A., & Amponsah, O. (2021). Climate change resilience: Lessons from local climate-smart agricultural practices in Ghana. *Energy, Ecology and Environment, 6*, 271–284.

Mohammed, K., Batung, E., Kansanga, M., Nyantakyi-Frimpong, H., & Luginaah, I. (2021). Livelihood diversification strategies and resilience to climate change in semi-arid northern Ghana. *Climatic Change, 164*, 1–23.

Muccione, V., Haasnoot, M., Alexander, P., Bednar-Friedl, B., Biesbroek, R., Georgopoulou, E., & Schmidt, D. N. (2024). Adaptation pathways for effective responses to climate change risks. *Wiley Interdisciplinary Reviews: Climate Change, 15*(4), Article e883.

Murer, C., & Piccoli, A. (2022). Affirmative policy in Nepal's community forestry: Does it make a difference in terms of social sustainability? *Sustainability, 14*(9), 5598.

Muringai, R. T., Mafongoya, P., & Lottering, R. T. (2022). Sub-Saharan Africa freshwater fisheries under climate change: A review of impacts, adaptation, and mitigation measures. *Fishes, 7*(3), 131.

Natarajan, N., Newsham, A., Rigg, J., & Suhardiman, D. (2022). A sustainable livelihoods framework for the 21st century. *World Development, 155*, Article 105898.

Ngcamu, B. S. (2023). Climate change effects on vulnerable populations in the Global South: A systematic review. *Natural Hazards, 118*(2), 977–991.

Obura, D., Gudka, M., Samoilys, M., Osuka, K., Mbugua, J., Keith, D. A., & Zivane, F. (2022). Vulnerability to collapse of coral reef ecosystems in the Western Indian Ocean. *Nature Sustainability, 5*(2), 104–113.

Okoronkwo, D. J., Ozioko, R. I., Ugwoke, R. U., Nwagbo, U. V., Nwobodo, C., Ugwu, C. H., & Mbah, E. C. (2024). Climate smart agriculture? Adaptation strategies of traditional agriculture to climate change in sub-Saharan Africa. *Frontiers in Climate, 6*, 1272320.

Omotoso, A. B., Letsoalo, S., Olagunju, K. O., Tshwene, C. S., & Omotayo, A. O. (2023). Climate change and variability in sub-Saharan Africa: A systematic review of trends and impacts on agriculture. *Journal of Cleaner Production, 414*, Article 137487.

Owolabi, O. P. (2023). *Ruling emancipated slaves and indigenous subjects: The divergent legacies of forced settlement and colonial occupation in the Global South*. Oxford University Press.

Pal, S. (2025). Rethinking resilience and vulnerability in adaptation studies. *Südasien-Chronik-South Asia Chronicle, 14*, 251–276.

Patnaik, H. A. (2021). Gender and participation in community based adaptation: Evidence from the decentralized climate funds project in Senegal. *World Development, 142*, Article 105448.

Pill, M. (2022). Towards a funding mechanism for loss and damage from climate change impacts. *Climate Risk Management, 35*, Article 100391.

Pinheiro, C. (2024). From the third world to the Global South: Definitions of moral geographies of inequality in anti-colonial intellectual traditions. *Sociology Compass, 18*(8), Article e13262.

Quandt, A., & Paderes, P. (2023). Livelihood resilience and global environmental change: Toward integration of objective and subjective approaches of analysis. *Geographical Review, 113*(4), 536–553.

Raj, S., Roodbar, S., Brinkley, C., & Wolfe, D. W. (2022). Food security and climate change: Differences in impacts and adaptation strategies for rural communities in the global south and north. *Frontiers in Sustainable Food Systems, 5*, Article 691191.

Robinne, F. N., & Secretariat, F. (2021). Impacts of disasters on forests, in particular forest fires. *UNFFS Background paper*.

Santpoort, R., Steel, G., Mkandawire, A., Ntauazi, C., Faye, E. H., & Githuku, F. (2021). The land is ours: Bottom-up strategies to secure rural women's access, control and rights to land in Kenya, Mozambique, Senegal and Malawi. *Frontiers in Sustainable Food Systems, 5*, Article 697314.

Shahin, M., Khanam, M., Aktar, S., Siddiqua, A., & Sharif, N. (2022). Resiliency of livelihood and empowerment of women: Results of a cash-based intervention in Bangladesh's Lalmonirhat District. *International Journal of Disaster Risk Reduction, 79*, Article 103137.

Shammin, M., Wang, A., & Sosland, M. (2022). A survey of community-based adaptation in developing countries. *Climate Change and Community Resilience*, 31.

Skovgaard, J., Adams, K. M., Dupuy, K., Dzebo, A., Funder, M., Fejerskov, A. M., & Shawoo, Z. (2023). Multilateral climate finance coordination: Politics and depoliticization in practice. *Global Environmental Politics, 23*(2), 125–147.

Songwe, V., Stern, N., & Bhattacharya, A. (2022). *Finance for climate action: Scaling up investment for climate and development*. Grantham Research Institute on Climate Change and the Environment, London School of Economics and Political Science.

Sulieman, H. M., & Young, H. (2023). The resilience and adaptation of pastoralist livestock mobility in a protracted conflict setting: West Darfur, Sudan. *Nomadic Peoples, 27*(1), 3–31.

Tefera, M. L., Seddaiu, G., Carletti, A., & Awada, H. (2025). Rainfall variability and drought in West Africa: Challenges and implications for rainfed agriculture. *Theoretical and Applied Climatology, 156*(1), 1–24.

Vertigans, S. (2021). CSR in the Global South: The continuing impact of post-colonial power and knowledge. In *Global challenges to CSR and sustainable development: Root causes and evidence from case studies* (pp. 67–85).

Ziervogel, G., Enqvist, J., Metelerkamp, L., & van Breda, J. (2022). Supporting transformative climate adaptation: Community-level capacity building and knowledge co-creation in South Africa. *Climate Policy, 22*(5), 607–622.

CHAPTER 2

Climate Resilience in Forest-Dependent Communities in the Developing World: What Implications for Livelihoods and Sustainability?

Abstract Forests play a vital role in supporting the livelihoods of communities in the Global South, providing essential resources such as food, income, medicinal plants, and vital ecosystem services. This chapter explores the interplay between forests, climate change, and livelihoods in the Global South, emphasizing the importance of local knowledge in building resilience. It equally highlights the critical role of policy in supporting adaptive strategies that empower communities and encourage sustainable forest management practices. Findings reveal that, these communities are particularly vulnerable to the impacts of climate change, which threaten forest ecosystems and the people who depend on them. Climate change-induced challenges, including altered rainfall patterns, droughts, pests, and biodiversity loss, significantly affect forest-dependent communities, exacerbating poverty and food insecurity. However, adaptive strategies like agroforestry, sustainable logging, and community-based conservation programs offer potential pathways for resilience. These strategies not only mitigate climate risks but also foster long-term livelihood sustainability by ensuring the health of forest ecosystems and promoting socio-economic stability. The integration of community-driven solutions and the promotion of equitable policies are necessary to safeguard both livelihoods and environmental sustainability. Through the prioritization of local involvement and adapting practices to

N. P. Awazi, *Building Resilience*, Palgrave Studies in Climate Resilient Societies, https://doi.org/10.1007/978-3-031-93791-0_2

changing climate conditions, forest-dependent communities can enhance resilience, protect their ecosystems, and achieve long-term sustainability.

Keywords Forest-dependent communities · Climate resilience · Livelihood sustainability · Global South · Agroforestry · Sustainable logging · Policy implications

2.1 Importance of Forests for Livelihoods in the Global South

Forests are crucial to the livelihoods of people in the Global South, where they provide a wide range of vital resources, including food, income, medicinal plants, and ecosystem services (Table 2.1). In these regions, particularly in low-income and rural communities, forests serve as a safety net and an essential source of sustenance, offering benefits that go beyond immediate survival. The intricate relationship between forests and the livelihoods of people in these regions is rooted in both the tangible and intangible services that forests provide.

In the Global South, forests are an integral source of food for millions of people, especially in rural areas. The forest provides a variety of foods, from fruits, nuts, and berries to bushmeat, mushrooms, and edible leaves. Many communities depend on forest-based food sources to meet their nutritional needs, especially in areas where agriculture may be limited by factors such as soil infertility, climate change, or natural disasters. According to Adhikari et al. (2016) and Ickowitz et al. (2022), forests contribute significantly to food security by providing resources that are easily accessible and often available during the lean seasons when other food supplies are scarce. In addition to direct consumption, forests also support agricultural productivity. For example, agroforestry systems, which combine trees with crops, are widely practiced in many parts of the Global South, especially in countries like Indonesia, Kenya, and India. These systems offer a diversified food supply, increase soil fertility, and protect against soil erosion. The inclusion of trees in farming systems has been shown to improve crop yields and reduce the vulnerability of farmers to climate-induced risks such as droughts or floods (Tebkew et al., 2024).

Table 2.1 The role of forests for livelihoods in the Global South

Importance of forests	*Description*	*Key references*
Food security	Forests provide various food resources such as fruits, nuts, berries, bushmeat, mushrooms, and edible leaves, which are crucial for nutritional needs, particularly in areas where agriculture is limited. They help ensure food security, especially during lean seasons	Adhikari et al. (2016), Ickowitz et al. (2022)
Support to agriculture	Forests support agricultural productivity through agroforestry systems that combine trees with crops, increasing soil fertility, protecting against erosion, and improving crop yields. These systems help buffer against climate-induced risks like droughts and floods	Tebkew et al. (2024)
Income generation	Forests provide income through timber, non-timber forest products (NTFPs), charcoal production, and ecotourism. NTFPs like honey, resins, and medicinal plants are vital for rural incomes, and the global trade in forest products supports economies in countries such as Brazil, Indonesia, and the Democratic Republic of Congo	Derebe et al. (2023), Rawat and Tekleyohannes (2021)
Medicinal plants	Forests offer medicinal plants used for traditional medicine, which is vital in countries like India, China, and Brazil. Many important pharmaceutical drugs have been derived from forest-based plants. However, overexploitation and climate change threaten the availability of these resources	WHO (2022)

(continued)

Table 2.1 (continued)

Importance of forests	*Description*	*Key references*
Ecosystem services	Forests provide essential ecosystem services, such as climate regulation, water regulation, soil protection, and clean air. They help mitigate climate change by storing carbon, regulate water cycles, and promote soil fertility, crucial for agriculture. These services are vital for communities in countries like Kenya, Nepal, and Brazil	Feng et al. (2022), Boscolo et al. (2021), Bhatt et al. (2024)
Cultural and spiritual significance	Forests have deep cultural and spiritual importance for many indigenous communities, serving as sacred spaces central to their identity and traditions. The destruction of forests threatens not only economic and ecological well-being but also cultural heritage	Bhatt et al. (2024)

Beyond providing food, forests in the Global South are crucial for generating income. Many people living in rural and forested areas rely on forest resources as a primary or supplementary source of livelihood. This income is derived from various activities such as the collection and sale of timber, non-timber forest products (NTFPs), charcoal production, and ecotourism. NTFPs, including fruits, nuts, honey, resins, and medicinal plants, represent an important and diverse income source for millions of people. According to Derebe et al. (2023), the collection and sale of NTFPs provide essential income for rural communities, particularly in parts of Africa, Asia, and Latin America. The global trade in forest products is also significant for economic growth in the Global South. In countries such as Brazil, Indonesia, and the Democratic Republic of Congo, timber and other forest products contribute substantially to national economies. In these regions, the forest sector provides direct and indirect employment, creating opportunities for both skilled and unskilled labor. However, the income derived from forests is not evenly distributed. While large commercial enterprises benefit from the extraction of timber and other forest resources, local communities, often in remote areas, depend on more sustainable forms of forest management, such as community forestry, to ensure equitable benefits (Rawat & Tekleyohannes, 2021). Furthermore, the forest-based income opportunities in the Global South are critical in providing resilience against economic shocks. Many households in forest-dependent regions rely on forest resources as a buffer during times of economic downturn or when other forms of income become unreliable. This is particularly important in areas where other livelihood options are limited due to underdevelopment, poor infrastructure, or lack of access to markets.

Forests in the Global South also play a vital role in providing medicinal plants, which have been used by indigenous and local communities for centuries to treat various ailments. Traditional medicine, which often relies on plants, fungi, and minerals found in forests, is an important part of healthcare systems in many parts of the Global South. In countries like India, China, and Brazil, herbal medicine is commonly practiced, and a significant proportion of the population continues to rely on forest resources for medicinal purposes. The diversity of plant species found in forests is an important factor in the development of traditional medicine. The World Health Organization (WHO) estimates that around 80% of the population in developing countries still relies on traditional medicine, much of which is derived from plants (WHO, 2022). In addition to local and national benefits, medicinal plants from forests in the Global South have increasingly gained global significance. The pharmaceutical industry often looks to these plants for the development of new drugs, with numerous important medications having been derived from forest-based plants. For instance, quinine, extracted from the bark of the cinchona tree found in the Amazon rainforest, has been a critical treatment for malaria. However, the overexploitation of forest resources, coupled with habitat destruction and climate change, poses a threat to the availability of medicinal plants. The loss of biodiversity in forest ecosystems could have severe consequences for local communities who depend on these plants for their health and well-being. Efforts to promote sustainable harvesting practices and protect indigenous knowledge related to medicinal plants are crucial for maintaining these important resources.

Forests offer a wide range of ecosystem services that are essential for the livelihoods of people in the Global South. These services include regulating services such as climate control, water regulation, soil protection, and the provision of clean air. Forests play a critical role in mitigating the effects of climate change by absorbing and storing carbon, thereby helping to reduce global warming. According to a study by Feng et al. (2022), tropical forests are responsible for storing approximately 25% of the world's terrestrial carbon. In the Global South, where many tropical and subtropical forests are located, these forests act as significant carbon sinks and contribute to global climate stability. Moreover, forests help regulate the water cycle by maintaining watersheds, preventing floods, and ensuring a stable water supply for communities (Boscolo et al., 2021). In countries like Kenya, Nepal, and Brazil, forests are vital for ensuring the availability of water for agriculture, drinking, and sanitation. They also contribute to soil fertility by preventing erosion and promoting nutrient cycling, which is essential for sustaining agricultural productivity. The loss of forests in these regions could lead to soil degradation, reduced agricultural yields, and increased vulnerability to natural disasters, further exacerbating poverty and food insecurity. Forest ecosystems also provide important cultural and spiritual services (Bhatt et al., 2024). Many indigenous communities in the Global South have deep cultural connections to forests, considering them sacred spaces that are central to their identity, spirituality, and way of life. The destruction of forests, therefore, threatens not only the economic and ecological well-being of these communities but also their cultural heritage.

Forests are therefore indispensable for the livelihoods of millions of people in the Global South, providing essential resources that sustain both local economies and cultural practices. They are a critical source of food, income, and medicinal plants, and they offer invaluable ecosystem services that support human well-being. However, the pressures of deforestation, climate change, and unsustainable resource extraction pose significant challenges to the continued benefits that forests provide. Efforts to promote sustainable forest management, protect biodiversity, and support community-based forest initiatives are crucial to ensuring that forests continue to support the livelihoods of people in the Global South.

2.2 Climate Change Impacts on Forests and Forest-Dependent Communities in the Global South

Forests in the Global South provide essential services such as carbon sequestration, biodiversity preservation, water regulation, and sustenance for millions of people. However, climate change is altering the dynamics of these ecosystems, with significant consequences for both the environment and the livelihoods of forest-dependent communities (Table 2.2).

One of the most significant climate change threats to forests in the Global South is the alteration of rainfall patterns. Many tropical forests, such as those in the Amazon, the Congo Basin, and Southeast Asia, rely heavily on consistent rainfall regimes to maintain their ecological balance. Climate models predict that climate change will result in more erratic rainfall, with some areas experiencing more intense rainfall events, while others face prolonged dry periods (Ngavom et al., 2024). In regions such as Sub-Saharan Africa, rainfall variability is already a source of concern for agricultural and forest-dependent communities. Studies show that altered precipitation patterns can lead to both flooding and drought, making it increasingly difficult for communities to adapt (Lombe et al., 2024). For instance, the Sahel region, which stretches across northern Africa, has seen a significant decrease in rainfall since the 1970s, contributing to desertification and the loss of critical forests (Yang et al., 2022). Such changes in rainfall patterns disrupt the growth cycles of trees, particularly those that are adapted to specific rainfall thresholds, leading to forest degradation and the weakening of ecosystem services like carbon storage and water regulation.

Increased frequency and intensity of droughts are another direct impact of climate change, particularly in arid and semi-arid regions of the Global South. Droughts place considerable stress on forests, affecting their survival and growth. As temperatures rise, the demand for water increases, and lower rainfall exacerbates the water deficit, which is particularly detrimental to tropical and subtropical forests that rely on regular moisture. The Amazon rainforest, often referred to as the "lungs of the planet," is already experiencing increasing frequency of droughts. A study by Wunderling et al. (2022) suggests that during drought years, the Amazon's ability to store carbon diminishes significantly, exacerbating the global warming problem. When forests experience drought stress, trees may die off, reducing the forest's capacity to regulate the climate and potentially triggering a feedback loop where the loss of forest cover accelerates climate change. This feedback mechanism is concerning, especially since forests in the Global South play a crucial role in mitigating climate change through carbon sequestration. For forest-dependent communities, droughts not only lead to the loss of forest resources, such as timber, fuelwood, and non-timber forest products (NTFPs), but they also diminish the availability of water for domestic use and agriculture. In countries

Table 2.2 Climate change impacts on forests and forest-dependent communities in the Global South

Climate change impact	*Description*	*Key references*
Altered rainfall patterns	Climate change is causing erratic rainfall, with some regions experiencing more intense rainfall while others face prolonged dry spells. This variability leads to flooding, drought, and desertification, particularly in regions like Sub-Saharan Africa and the Sahel, impacting forest growth and ecosystem services	Ngavom et al. (2024), Lombe et al. (2024), Yang et al. (2022)
Increased frequency of droughts	Rising temperatures and lower rainfall intensify droughts, particularly in arid and semi-arid regions. Droughts harm tropical forests like the Amazon, reducing their ability to store carbon, which exacerbates global warming. Forest-dependent communities also face water shortages, food insecurity, and income loss due to droughts	Wunderling et al. (2022), Algur et al. (2021), Dongmo (2022)
Forest pests and diseases	Warmer temperatures and changing precipitation patterns lead to the spread of forest pests and diseases. Insects like the coffee borer beetle and cassava mealybug thrive in warmer conditions, causing extensive damage to forests and crops, undermining the livelihoods of forest-dependent communities	Gullino et al. (2022), Subedi et al. (2023), Parthasarathy et al. (2025)
Loss of biodiversity	Climate change disrupts species' life cycles and migration patterns, leading to shifts in species composition and loss of biodiversity. This threatens the availability of food, medicine, and income resources for forest-dependent communities, while diminishing ecosystem resilience to climate change	Bogoni et al. (2023), Beekmann et al. (2024), Muluneh (2021)

(continued)

Table 2.2 (continued)

Climate change impact	*Description*	*Key references*
Shifting forest boundaries	As temperatures rise, tropical forests are moving to higher altitudes and latitudes, while dry forests may transform into grasslands. These shifts affect carbon sequestration, water availability, and forest-based resources, which in turn impact the livelihoods of forest-dependent communities	Elsen et al. (2022), Jaramillo (2023), Arizapana-Almonacid et al. (2024)
Socio-economic inequalities	Climate change exacerbates socio-economic inequalities, with marginalized forest-dependent communities facing the most severe impacts. Indigenous peoples and local farmers, who rely on forests for resources and cultural practices, are vulnerable to displacement, loss of resources, and conflicts driven by environmental changes	Felipe Pérez and Tomaselli (2021), Reyes-García et al. (2024)
Impacts on forest ecosystem services	The combination of altered rainfall, drought, pests, and shifting forest boundaries threatens essential ecosystem services, such as carbon sequestration, water regulation, and soil protection. These disruptions reduce the forests' ability to mitigate climate change and support local economies	Algur et al. (2021), Wunderling et al. (2022), Muluneh (2021)

like India and Kenya, prolonged droughts threaten the livelihoods of millions of people who depend on forest ecosystems for water, food, and income (Algur et al., 2021; Dongmo, 2022). These communities often lack the resources to cope with such extreme events, which can lead to displacement, food insecurity, and social conflicts.

The changing climate also affects the dynamics of forest pests and diseases, which are becoming more prevalent and severe in response to warmer temperatures and altered precipitation patterns. Insects such as the mountain pine beetle in North America or the coffee borer beetle in Central America have been able to expand their ranges due to warmer winters, leading to significant forest damage (Gullino et al., 2022; Subedi et al., 2023). Similarly, in the Global South, warmer conditions favor the spread of pests and pathogens that weaken trees and forest ecosystems. In Southeast Asia, for example, increased temperatures and humidity have facilitated the spread of the cassava mealybug and the coconut rhinoceros beetle, both of which have caused extensive damage to forests and crops (Parthasarathy et al., 2025). These pests not only degrade forest health but also undermine the livelihoods of communities that depend on forest products, such as timber, fruits, and medicinal plants. As the climate continues to warm, the frequency and severity of pest outbreaks are likely to increase, compounding the vulnerability of both forests and forest-dependent people.

Forests in the Global South are biodiversity hotspots, hosting a vast array of plant and animal species. However, climate change is significantly impacting biodiversity, threatening the survival of species that are already vulnerable due to habitat loss, overexploitation, and other anthropogenic pressures. Changes in temperature and precipitation patterns are disrupting species' life cycles and migration patterns, leading to shifts in species composition and the fragmentation of habitats. In the Amazon, for example, climate change has been linked to the degradation of forest ecosystems and the loss of species, including iconic species like jaguars and the harpy eagle (Bogoni et al., 2023). Similarly, in Africa's Congo Basin, rising temperatures and reduced rainfall are driving shifts in species distributions, while increased forest fragmentation from human activities exacerbates the effects (Beekmann et al., 2024). For forest-dependent communities, the loss of biodiversity represents a direct threat to their livelihoods. Many communities rely on the diverse flora and fauna of forests for food, medicine, and income. The decline in key species of plants, animals, and insects can reduce the availability of these essential resources. Moreover, as biodiversity declines, the resilience of ecosystems to climate change is diminished, further exacerbating the impacts of changing weather patterns and increasing the vulnerability of these communities (Muluneh, 2021).

As temperatures rise and precipitation patterns change, the boundaries of different forest types are also shifting. Tropical forests are migrating toward higher altitudes and latitudes, while some dry forests may convert into savannas or grasslands (Elsen et al., 2022; Jaramillo, 2023). Such shifts in forest distribution have significant implications for the ecosystem services that forests provide. For example, as forests shrink or shift, carbon sequestration potential diminishes, contributing to further warming. Forest-dependent communities are particularly vulnerable to these shifts because they may no longer have access to the same resources or ecosystem services they once relied on. For example, indigenous communities in the Andes are experiencing changing water availability as higher-altitude forests are replaced by scrublands due to warmer temperatures (Arizapana-Almonacid et al., 2024). This reduces the availability of water and other vital forest products, impacting local economies and ways of life.

In addition to environmental changes, climate change also exacerbates socio-economic inequalities, with marginalized communities in the Global South bearing the brunt of its impacts. Indigenous peoples, local farmers, and forest-dependent communities often lack the resources to adapt to the changes brought about by climate change. These communities rely on forests for food, medicine, building materials, and cultural practices, and the degradation of these ecosystems poses direct threats to their survival. The encroachment of industrial agriculture, mining, and logging into forested areas further compounds these risks. The increasing competition for land and resources often leads to conflicts, displacement, and the erosion of traditional knowledge and practices. Studies by Felipe Pérez and Tomaselli (2021) and Reyes-García et al. (2024) found that indigenous communities in Central America are increasingly vulnerable to the impacts of climate change, as they depend on forests for much of their income and cultural heritage.

The impacts of climate change on forests and forest-dependent communities in the Global South are profound and multifaceted. Altered rainfall patterns, droughts, pests, biodiversity loss, and the shifting of forest boundaries all present significant challenges for these ecosystems and the millions of people who rely on them. The complex interaction between these factors not only threatens the environment but also exacerbates social and economic vulnerabilities, particularly for marginalized communities. To address these challenges, a multifaceted approach is needed, including enhanced climate adaptation strategies, the protection of biodiversity, and the empowerment of local communities to build resilience to climate impacts. The importance of forests in regulating the global climate and supporting livelihoods cannot be overstated, and efforts to mitigate and adapt to climate change must prioritize the protection and sustainable management of forests in the Global South.

2.3 Adaptive Strategies in Forest Communities in the Global South

Forest communities in the Global South face the dual challenge of preserving their ecosystems while achieving sustainable livelihoods. The pressure from logging, agricultural expansion, and climate change poses significant threats to both forest health and the communities dependent on them. However, these communities have developed adaptive strategies, leveraging local knowledge, traditional practices, and innovations to maintain ecological balance and economic stability. Various adaptive strategies are being used by forest communities in the Global South, including agroforestry, sustainable logging, community-based conservation programs, and other strategies (Table 2.3).

Agroforestry, the practice of integrating trees into agricultural landscapes, has emerged as a key adaptive strategy in forest communities across the Global South. Through the combination of trees, crops, and livestock, agroforestry systems offer a way to diversify income, improve soil fertility, and enhance forest regeneration while reducing pressure on natural forests (Chappa et al., 2024; Vinodhini et al., 2023). These systems often follow traditional knowledge and practices, passed down through generations, but have also incorporated modern scientific knowledge to optimize productivity and sustainability. In the Philippines, the Department of Environment and Natural Resources (DENR) has implemented various agroforestry programs aimed at restoring degraded lands and providing alternative livelihoods for local communities. In the province of Quezon, farmers have been involved in the implementation of "rainforestation," an agroforestry practice that integrates native tree species with crops such as cacao, coconut, and abaca. This approach not only provides a steady income but also contributes to forest restoration and biodiversity conservation (Fernandez Jr. et al., 2021, 2024). The success of these agroforestry systems is linked to the knowledge and involvement of local communities, whose active participation ensures that interventions align with their needs and knowledge systems. While agroforestry is often hailed as a win–win solution for both conservation and livelihood, its success is contingent on supportive policies, access to markets, and community engagement. Many forest-dependent communities struggle with limited access to agricultural inputs, credit, and technical support (Mantey & Teye, 2021). Thus, for agroforestry to be effective in the Global South, it must be supported by government policies that facilitate

Table 2.3 Adaptive strategies used by forest communities in the Global South

Adaptive strategy	*Description*	*Examples*	*Key references*
Agroforestry	Integrating trees into agricultural landscapes to diversify income, improve soil fertility, and enhance forest regeneration. This reduces pressure on natural forests	Agroforestry programs in the Philippines, including "rainforestation" in Quezon province, integrating native tree species with crops like cacao, coconut, and abaca	Vinodhini et al. (2023), Chappa et al. (2024), Fernandez Jr. et al. (2021), Fernandez Jr. et al. (2024)
Sustainable logging	Balancing timber extraction with forest preservation. Sustainable logging practices reduce deforestation while maintaining forest health and providing income to local communities	Community Forest User Groups (CFUGs) in Nepal, such as the Barpak CFUG, implementing selective logging for forest regeneration and biodiversity preservation	Charmakar et al. (2024), Poudyal et al. (2022)
Community-Based Conservation (CBC)	Involves local communities in the management and protection of natural resources, aligning conservation with local development needs	The Shuar people in the Ecuadorian Amazon managing Protected Forests, integrating traditional knowledge and modern conservation science	Kor et al. (2021), Dawson et al. (2021)
Ecotourism	Using forest biodiversity to support sustainable tourism, providing income for local communities while promoting conservation	Costa Rica's ecotourism initiatives in the Monteverde region, where local farmers have shifted to nature tours and hospitality services	Miller et al. (2023)

(continued)

Table 2.3 (continued)

Adaptive strategy	*Description*	*Examples*	*Key references*
Fire management	Community-based programs to prevent and control forest fires, combining traditional knowledge with modern techniques	Central Kalimantan in Indonesia, where local communities manage forest fires using both traditional and modern fire control methods	Guevara Arce (2023)
Climate adaptation measures	Strategies to help communities adapt to the impacts of climate change, such as protecting forests and managing forest ecosystems	Programs aimed at restoring degraded lands and providing alternative livelihoods through agroforestry	Mantey and Teye (2021)

land tenure security, access to extension services, and equitable market opportunities.

Sustainable logging practices are another adaptive strategy employed by forest communities in the Global South. These practices seek to balance timber extraction with the preservation of forest ecosystems and the livelihoods of local people. Unlike conventional logging, which often leads to deforestation, sustainable logging aims to minimize environmental impacts while maintaining economic viability. Nepal provides a noteworthy example of sustainable logging through its Community Forest User Groups (CFUGs). These groups manage forests under the national forest policy, which grants local communities the right to use and protect forest resources. Since the 1990s, Nepal has decentralized forest management to local communities, who are responsible for sustainable timber extraction, non-timber forest products (NTFPs), and conservation efforts (Charmakar et al., 2024). The CFUGs have shown that community-based management can significantly reduce deforestation and degradation while maintaining forest health and generating income. In one case, the Barpak CFUG in Gorkha District has implemented sustainable logging practices, where selective logging is carried out based on forest health assessments. This approach has led to a steady income for the community while improving forest regeneration and biodiversity (Poudyal et al., 2022). Moreover, CFUGs have been successful in empowering marginalized groups, including women and indigenous peoples, by involving them in decision-making processes and ensuring that benefits are distributed equitably. Sustainable logging faces challenges such as limited technical capacity, market access for sustainably sourced timber, and pressure from commercial logging operations. Despite these obstacles, community-based management of forests is proving to be an effective strategy for both conserving biodiversity and generating local income. To scale this model, it is critical to invest in capacity-building, ensure fair market prices for sustainably sourced timber, and strengthen local governance structures.

Community-based conservation (CBC) programs have emerged as a central strategy in forest management across the Global South. These programs emphasize the active involvement of local communities in the management and protection of their natural resources. CBC programs aim to align conservation goals with local development needs, creating incentives for communities to engage in sustainable practices. In the Ecuadorian Amazon, the Shuar people have developed community-based conservation programs to protect their forests from external threats such as oil exploration and logging. The Shuar peoples run Protected Forests programs covering over 200,000 hectares of tropical rainforest. These programs involve sustainable use of forest resources, including the management of non-timber products such as nuts, fruits, and medicinal plants (Kor et al., 2021). The Shuar's deep cultural connection to the land and their traditional ecological knowledge have been key to the success of this program, as they have been able to integrate modern conservation science with their ancestral practices. While CBC programs offer promising outcomes, they also face challenges, including insufficient funding, external political pressures, and conflicts over land tenure. Moreover, the involvement of local communities in conservation often requires overcoming external actors' biases against indigenous knowledge and practices (Dawson et al., 2021). The success of CBC programs hinges on fostering partnerships between governments, NGOs, and local communities, ensuring that conservation efforts are both culturally appropriate and economically beneficial.

In addition to agroforestry, sustainable logging, and community-based conservation, other adaptive strategies have emerged in response to the challenges faced by forest communities in the Global South. These strategies include ecotourism, fire management, and climate adaptation measures. Costa Rica has become a global leader in ecotourism, using its rich biodiversity and forests as the basis for tourism development. In the

1980s, Costa Rica implemented a system of national parks and protected areas that involved local communities in conservation efforts. Ecotourism has become a vital source of income for forest-dependent communities, particularly in the region of Monteverde, where local farmers have shifted from traditional agriculture to offering nature tours and hospitality services (Miller et al., 2023). By integrating conservation with tourism, Costa Rica has been able to provide communities with financial incentives to protect their forests while educating tourists about environmental issues. In Indonesia, forest fires have become a major issue due to agricultural expansion and the illegal logging industry. However, community-based fire management programs have been implemented in regions such as Central Kalimantan, where local communities have been trained to prevent and control fires. These programs are based on traditional fire management practices and have been successful in reducing the frequency and severity of forest fires (Guevara Arce, 2023). Through a combination of traditional knowledge with modern fire control techniques, these initiatives help protect both the forests and the livelihoods of local communities.

The adaptive strategies employed by forest communities in the Global South illustrate the resilience and innovation of these communities in the face of environmental and socio-economic challenges. Agroforestry, sustainable logging, community-based conservation, and other adaptive strategies offer promising solutions to the complex problems of forest degradation, biodiversity loss, and livelihood insecurity. However, for these strategies to be successful in the long term, they must be supported by appropriate policies, investments in capacity-building, and the active involvement of local communities in decision-making processes.

2.4 Livelihood Sustainability in Forest-Dependent Communities in the Global South

Livelihood sustainability in forest-dependent communities in the Global South is a multifaceted issue that links environmental health, socio-economic stability, and cultural continuity. Forests play a vital role in providing resources for millions of people in these regions. However, as global pressures such as climate change, deforestation, and socio-economic changes affect the ability of these communities to sustain their livelihoods, adaptive strategies become key to ensuring long-term sustainability. These strategies help to maintain the delicate balance between meeting the immediate needs of the population, ensuring environmental resilience, and preserving socio-economic structures.

In the Global South, forests serve as a crucial resource for both rural and urban populations. They provide a variety of direct and indirect services, including food, fuel, medicinal plants, construction materials, and income from timber and non-timber forest products (NTFPs) (Amusa et al., 2024; Mugari et al., 2024). For example, in countries like India, Brazil, and Indonesia, forests contribute significantly to rural economies and are integral to the livelihoods of indigenous and local communities (Duffy et al., 2021; Gunawan et al., 2022). Forests also regulate the climate and maintain biodiversity, making them an essential part of the ecological health of the region. However, these communities face various challenges, including land tenure insecurity, insufficient access to markets, inadequate infrastructure, and environmental degradation. Many communities are also vulnerable to external shocks such as climate change, shifting market demands, and political instability (Axelsson & Martins, 2024). Consequently, maintaining livelihood sustainability requires adaptive strategies that are both flexible and resilient to these pressures.

In the context of forest-dependent communities, adaptive strategies can take various forms, ranging from the diversification of income sources to the adoption of sustainable land-use practices. These strategies contribute to long-term livelihood sustainability by enabling communities to cope with and thrive amid external pressures. One of the primary adaptive strategies for forest-dependent communities is the diversification of livelihood sources. In many cases, communities that traditionally relied on a single resource or activity, such as logging or agriculture, have found ways to diversify their income streams by integrating other forest products, such as honey, mushrooms, or medicinal plants, into their livelihoods (Asamoah et al., 2024). Diversification reduces dependency on a single income source and helps cushion against market fluctuations or environmental changes that might impact specific sectors. For instance, in the Amazon Basin, many indigenous communities have turned to ecotourism and agroforestry as supplementary income sources (Abramovay et al., 2021; Gunter & Ceddia, 2021). These activities not only provide additional income but also promote sustainable land-use practices that enhance the overall health of the ecosystem. Diversification is thus a key strategy for ensuring that these communities are not solely dependent on the state of any one market or resource, improving both ecological and economic resilience.

Sustainable forest management (SFM) is another adaptive strategy crucial for long-term livelihood sustainability. SFM includes a range of practices designed to balance the use of forest resources with the need to maintain ecosystem health over time (Condé et al., 2022). For example, selective logging, which involves removing only mature trees while leaving the younger ones intact, helps to maintain the forest's structural integrity and biodiversity, ensuring that the forest continues to provide ecosystem services (Borma et al., 2022). Agroforestry systems, where agricultural crops are integrated with tree planting, also represent a form of sustainable management that benefits both the environment and the livelihoods of forest-dependent communities. In places like Sub-Saharan Africa and Southeast Asia, agroforestry systems have been found to increase soil fertility, reduce erosion, and improve the availability of forest products, while also providing communities with more diverse income opportunities (Castle et al., 2021; Muthee et al., 2022). These practices are essential for maintaining the long-term health of forests while also meeting the needs of the local population.

Indigenous and local knowledge systems play a pivotal role in adaptive strategies for forest-dependent communities. These knowledge systems, often passed down through generations, include insights into sustainable land use, resource management, and biodiversity conservation (Akalibey et al., 2024; Gordon et al., 2023). For example, indigenous communities in the Amazon have developed sophisticated systems for managing fish populations, collecting non-timber forest products, and protecting sacred forest areas from overexploitation (Palma Valente, 2022). Incorporating traditional ecological knowledge (TEK) into modern conservation and development programs can enhance the effectiveness of adaptive strategies by ensuring that they are culturally appropriate and context-specific. Moreover, the revitalization of indigenous practices can foster a sense of cultural pride and social cohesion, which is essential for the long-term sustainability of these communities (Bawi, 2025).

The health of ecosystems is directly tied to the sustainability of the livelihoods of forest-dependent communities (Ela et al., 2021). Forests provide a wide range of ecosystem services, such as carbon sequestration, water regulation, and biodiversity conservation. When ecosystems are degraded, the availability of resources for local communities diminishes, and the social and economic stability of these communities is compromised. Adaptive strategies that focus on ecosystem restoration and protection contribute significantly to long-term livelihood sustainability by ensuring that ecosystems remain functional and resilient. For instance, the restoration of degraded forest lands through reforestation and afforestation projects can help to recover biodiversity, improve soil fertility, and mitigate the impacts of climate change. In Tanzania, for example, community-led forest management initiatives have successfully restored degraded lands, increasing forest cover and providing new income opportunities for local people through sustainable harvesting practices (Noe et al., 2022). Moreover, adaptive strategies that focus on ecosystem health also benefit communities by reducing the risks associated with environmental hazards. Forests, when well-managed, can act as buffers against natural disasters such as floods, landslides, and droughts. In regions like Southeast Asia, community-based forest management has been shown to improve resilience to climate change impacts, such as increased flooding and changing rainfall patterns (Karki et al., 2022; Tota et al., 2024).

In addition to environmental benefits, adaptive strategies also contribute to socio-economic stability in forest-dependent communities. When communities adopt sustainable practices and diversify their livelihoods, they reduce their vulnerability to economic shocks and improve their overall well-being. For example, in the Philippines, community-based forest management programs have led to the establishment of sustainable income-generating activities, such as timber harvesting, NTFP collection, and ecotourism, all of which provide communities with stable and diverse income streams (Pulhin et al., 2024). These initiatives also promote greater social cohesion by fostering collective decision-making and shared ownership of resources. Moreover, adaptive strategies that integrate gender equality and social inclusion are essential for fostering socio-economic stability. Empowering women and marginalized groups in forest-dependent communities through access to resources, education, and decision-making processes can significantly enhance the overall success of livelihood sustainability efforts (Phuong et al., 2024). Gender-inclusive strategies have been shown to improve the resilience of communities by ensuring that all members have the resources and capabilities to adapt to changes in their environment and economy.

Adaptive strategies therefore play a critical role in ensuring the long-term sustainability of livelihoods in forest-dependent communities in the Global South. Through the diversification of income sources, adopting sustainable land-use practices, and incorporating traditional ecological knowledge, these communities can better cope with environmental, economic, and social challenges. At the same time, these strategies help to maintain ecosystem health, which in turn supports the stability of local economies and cultures. The integration of adaptive strategies into forest management policies and practices is not only crucial for the well-being of these communities but also for the broader goal of sustainable development. Promoting adaptive strategies that prioritize ecosystem resilience and socio-economic stability is essential for ensuring the sustainability of livelihoods in forest-dependent communities in the Global South.

2.5 Policy Implications

The intersection of climate resilience, forest-dependent communities, and sustainability is critical in shaping the future of rural populations in developing countries. Forest-dependent communities, particularly in regions vulnerable to climate change, rely heavily on forests for their livelihoods. These communities have long managed forests and ecosystems in ways that ensure their survival, maintaining a delicate balance between their economic needs and the environmental health of the land. However, with the rising impacts of climate change such as unpredictable weather patterns, increased frequency of natural disasters, and shifting biodiversity, there is an urgent need to reassess current policy frameworks. To address these challenges, policies must focus on integrating community knowledge, empowering local people, and promoting sustainable forest management practices.

2.5.1 *Integration of Community Knowledge into Policy Development*

One of the most important aspects of fostering climate resilience in forest-dependent communities is the integration of local and indigenous knowledge into formal policy-making processes. Many of these communities have an intimate understanding of the local environment, gained through generations of experience. This knowledge includes insights into biodiversity, water cycles, seasonal changes, and the resilience of different forest species. However, this valuable resource often goes unrecognized or is underutilized in policy development, which can undermine the sustainability of interventions and the overall resilience of the community. Policymakers need to prioritize collaborative decision-making frameworks that respect and incorporate the experiences of forest communities. This means engaging these communities in the design and implementation of climate resilience strategies. Community-led conservation efforts, sustainable agricultural practices, and forest restoration initiatives have demonstrated that local knowledge can greatly enhance the success of climate adaptation strategies. Furthermore, policies should include mechanisms for formalizing this knowledge, ensuring it is integrated into climate resilience plans, and recognized as a key element of sustainable forest management.

2.5.2 *Empowering Local People Through Capacity-Building and Education*

For policies to be truly effective in enhancing climate resilience, it is essential that they empower local communities. One of the critical ways to achieve this is through capacity-building, education, and skill development. Forest-dependent communities often face significant challenges in terms of access to information, technical skills, and financial resources. These barriers can hinder their ability to engage with modern technologies, apply climate-smart practices, or benefit from governmental and international climate adaptation programs. Policymakers must invest in educational programs that are tailored to the specific needs and contexts of forest-dependent communities. These programs should focus on climate change awareness, sustainable forest management techniques, alternative livelihoods, and the importance of biodiversity conservation. Additionally, efforts should be made to provide training in the use of digital tools and modern technologies, allowing communities to access weather forecasting, satellite data, and other climate-related information that can help them better anticipate and respond to climate shocks. Moreover, policies should foster partnerships with NGOs, research institutions, and local governments to create knowledge-sharing platforms. By building local expertise, communities will be better equipped to make informed decisions regarding forest resource management, which can lead to more sustainable and resilient livelihoods.

2.5.3 Promoting Sustainable Forest Management Practices

Sustainable forest management (SFM) is essential for the long-term resilience of both the ecosystems and the communities that depend on them. Effective SFM can mitigate the adverse impacts of climate change, such as soil erosion, forest degradation, and biodiversity loss, while also providing a steady source of income for communities through non-timber forest products, ecotourism, and sustainable logging. Policies should encourage the adoption of sustainable forest management practices by providing the necessary incentives and support. For instance, financial mechanisms such as payments for ecosystem services (PES), carbon credits, or subsidies for sustainable harvesting methods could be implemented to encourage forest conservation and restoration. These mechanisms can help communities maintain their forests while also generating income, ensuring that forest conservation is seen not only as a duty but also as a livelihood strategy. Governments and international organizations should support the development of certification systems that recognize sustainable forest management practices. For instance, the Forest Stewardship Council (FSC) certification offers a globally recognized standard that can increase the value of forest products from sustainably managed forests. Additionally, policies should enable access to markets for certified forest products, thereby ensuring that communities can benefit financially from sustainable practices. Furthermore, policies should aim to reduce the pressure on forests from unsustainable practices, such as illegal logging and land conversion for agriculture. By providing alternative income-generating activities, such as agroforestry, sustainable ecotourism, and community-based forest enterprises, policymakers can reduce communities' reliance on harmful practices and foster a shift toward sustainability.

2.5.4 Climate Adaptation Financing and Access to Resources

The implementation of climate resilience strategies requires adequate funding and resource allocation. Developing countries, especially those with high levels of poverty, often face challenges in accessing climate adaptation funds, such as those offered through the Green Climate Fund or other international climate financing mechanisms. Policies should ensure that forest-dependent communities have the capacity to access these funds and are included in national and international climate finance mechanisms. To facilitate this, governments should work toward creating transparent and accessible channels for financing climate adaptation projects at the community level. These channels should allow local communities to access funds directly, reducing reliance on intermediaries that may not fully understand the unique needs of forest-dependent populations. Furthermore, policies should provide financial literacy training to these communities, enabling them to manage funds effectively and ensure that investments are directed toward projects that will enhance their climate resilience.

2.5.5 *Strengthening Institutional and Governance Frameworks*

Strong institutional frameworks and governance structures are essential for the effective implementation of policies aimed at enhancing climate resilience in forest-dependent communities. Policymakers must ensure that local governance structures are equipped to address the challenges posed by climate change. This includes creating local committees or advisory boards that can serve as a bridge between national policies and community needs, ensuring that decisions reflect the realities on the ground. Additionally, the role of women and marginalized groups in forest-dependent communities must be highlighted in policy frameworks. These groups often have valuable knowledge and unique perspectives on sustainable resource management. Empowering them through inclusive policies will not only promote gender equality but also enhance the overall resilience of the community.

References

Abramovay, R., Ferreira, J., Costa, F. D. A., Ehrlich, M., Euler, A. M. C., Young, C. E. F., & Villanova, L. (2021). The new bioeconomy in the Amazon: Opportunities and challenges for a healthy standing forest and flowing rivers. *Amazon Assessment Report 2021.*

Adhikari, J., Ojha, H., & Bhattarai, B. (2016). Edible forest? Rethinking Nepal's forest governance in the era of food insecurity. *International Forestry Review, 18*(3), 265–279.

Akalibey, S., Hlaváčková, P., Schneider, J., Fialová, J., Darkwah, S., & Ahenkan, A. (2024). Integrating indigenous knowledge and culture in sustainable forest management via global environmental policies. *Journal of Forest Science, 70*(6), 265.

Algur, K. D., Patel, S. K., & Chauhan, S. (2021). The impact of drought on the health and livelihoods of women and children in India: A systematic review. *Children and Youth Services Review, 122*, Article 105909.

Amusa, T. O., Avana-Tientcheu, M. L., Awazi, N. P., & Chirwa, P. W. (2024). The role of non-timber forest products for sustainable livelihoods in African multifunctional landscapes. In *Trees in a Sub-Saharan multi-functional landscape: Research, management, and policy* (pp. 153–178). Springer Nature Switzerland.

Arizapana-Almonacid, M. A., Pariona-Antonio, V. H., Castañeda-Tinco, I., Ascención Mendoza, J. C., Gutiérrez Gómez, E., & Ramoni-Perazzi, P. (2024). Land cover changes and comparison of current landscape metrics in a region of the Central Andes affected by population migration. *Annals of GIS, 30*(1), 105–120.

Asamoah, O., Danquah, J. A., Bamwesigye, D., Boakye, E. A., Appiah, M., & Pappinen, A. (2024). Perception of locals on multiple contributions of NTFPs to the livelihoods of forest fringe communities in Ghana. *Forests, 15*(5), 861.

Axelsson, T., & Martins, I. (2024). Resilience to shrinking as a catch-up strategy: A comparison of Brazil and Indonesia, 1964–2019. *Studies in Comparative International Development, 59*(3), 491–516.

Bawi, S. V. (2025). Revitalizing ritual performances for environmental conservation and climate change adaptation: A case study of ecotourism in Khaw Nu M'cung National Park, Myanmar. *Forest and Society, 9*(1), 48–69.

Beekmann, M., Gallois, S., & Rondinini, C. (2024). Uncertain future for Congo Basin biodiversity: A systematic review of climate change impacts. *Biological Conservation, 297*, Article 110730.

Bhatt, H., Jugran, H. P., & Pandey, R. (2024). Cultural ecosystem services nexus with socio-cultural attributes and traditional ecological knowledge for managing community forests of Indian western Himalaya. *Ecological Indicators, 166*, Article 112379.

Bogoni, J. A., Boron, V., Peres, C. A., Coelho, M. E. M., Morato, R. G., & Oliveira-da-Costa, M. (2023). Impending anthropogenic threats and protected area prioritization for jaguars in the Brazilian Amazon. *Communications Biology, 6*(1), 132.

Borma, L. S., Costa, M. H., da Rocha, H. R., Arieira, J., Nascimento, N. C. C., Jaramillo-Giraldo, C., & Nobre, C. A. (2022). Beyond carbon: The contributions of South American tropical humid and subhumid forests to ecosystem services. *Reviews of Geophysics, 60*(4), e2021RG000766.

Boscolo, M., Leonardi, A., Masiero, M., Amato, G., Laghetto, G., & O'Driscoll, C. (2021). Valuing water from forests. In *A guide to forest–water management* (Vol. 185, pp. 75–106). FAO, IUFRO, USDA.

Castle, S. E., Miller, D. C., Ordonez, P. J., Baylis, K., & Hughes, K. (2021). The impacts of agroforestry interventions on agricultural productivity, ecosystem services, and human well-being in low-and middle-income countries: A systematic review. *Campbell Systematic Reviews, 17*(2), Article e1167.

Chappa, L. R., Nungula, E. Z., Makwinja, Y. H., Ranjan, S., Sow, S., Alnemari, A. M., & Gitari, H. I. (2024). Outlooks on major agroforestry systems. *Agroforestry*, 21–48.

Charmakar, S., Kimengsi, J. N., & Giessen, L. (2024). Linking institutional change mechanisms with forest management outcomes: Evidence from community forestry in Nepal. *Ecology and Society*, *29*(3).

Condé, T. M., Tonini, H., Higuchi, N., Higuchi, F. G., Lima, A. J. N., Barbosa, R. I., & Haas, M. A. (2022). Effects of sustainable forest management on tree diversity, timber volumes, and carbon stocks in an ecotone forest in the northern Brazilian Amazon. *Land Use Policy, 119*, Article 106145.

Dawson, N. M., Coolsaet, B., Sterling, E. J., Loveridge, R., Gross-Camp, N. D., Wongbusarakum, S., & Rosado-May, F. J. (2021). The role of Indigenous peoples and local communities in effective and equitable conservation. *Ecology and Society, 26*(3), 19.

Derebe, B., Alemu, A., & Asfaw, Z. (2023). Contribution of nontimber forest products earn to livelihood in rural households and the type of use: A systematic review. *International Journal of Forestry Research, 2023*(1), 9643290.

Dongmo, C. (2022). Resilience to environmental challenges and the national disaster insurance program in Kenya. In *Energy policy advancement: Climate change mitigation and international environmental justice* (pp. 145–161).

Duffy, C., Toth, G. G., Hagan, R. P., McKeown, P. C., Rahman, S. A., Widyaningsih, Y., & Spillane, C. (2021). Agroforestry contributions to smallholder farmer food security in Indonesia. *Agroforestry Systems, 95*(6), 1109–1124.

Ela, M. Z., Rabya, T., Khan, L., Rahman, M. H., Shovo, T. E. A., Jahan, N., & Islam, M. N. (2021). Climate change and livelihood vulnerabilities: The forest resource-dependent communities of the Sundarbans of Bangladesh. In *Climate vulnerability and resilience in the Global South: Human adaptations for sustainable futures* (pp. 341–352).

Elsen, P. R., Saxon, E. C., Simmons, B. A., Ward, M., Williams, B. A., Grantham, H. S., & Watson, J. E. (2022). Accelerated shifts in terrestrial life zones under rapid climate change. *Global Change Biology, 28*(3), 918–935.

Felipe Pérez, B., & Tomaselli, A. (2021). Indigenous peoples and climate-induced relocation in Latin America and the Caribbean: Managed retreat as a tool or a threat? *Journal of Environmental Studies and Sciences, 11*(3), 352–364.

Feng, Y., Zeng, Z., Searchinger, T. D., Ziegler, A. D., Wu, J., Wang, D., & Zheng, C. (2022). Doubling of annual forest carbon loss over the tropics during the early twenty-first century. *Nature Sustainability, 5*(5), 444–451.

Fernandez, G. C., Jr., Bande, M. J. M., Quebecc, J. C., & Pogosa, J. O. (2021). Rainforestation as an instrument for environmental education: A Philippine rural community experience. *Turkish Online Journal of Qualitative Inquiry*, *12*(7).

Fernandez, G. C., Jr., Pogosa, J. O., & Bande, M. M. (2024). The role of women in conceptualizing, promoting, and implementing rainforestation native tree forest restoration. *Social Ethics Society Journal of Applied Philosophy*, 273–295.

Gordon, H. S. J., Ross, J. A., Bauer-Armstrong, C., Moreno, M., Byington, R., & Bowman, N. (2023). Integrating indigenous traditional ecological knowledge of land into land management through indigenous-academic partnerships. *Land Use Policy, 125*, Article 106469.

Guevara Arce, S. M. (2023). Towards a fire safety framework in informal settlements in Costa Rica. https://hdl.handle.net/10315/41347

Gullino, M. L., Albajes, R., Al-Jboory, I., Angelotti, F., Chakraborty, S., Garrett, K. A., & Stephenson, T. (2022). Climate change and pathways used by pests as challenges to plant health in agriculture and forestry. *Sustainability, 14*(19), 12421.

Gunawan, H., Yeny, I., Karlina, E., Suharti, S., Murniati, S., & Nurlia, A. (2022). Integrating social forestry and biodiversity conservation in Indonesia. *Forests, 13*(12), 2152.

Gunter, U., & Ceddia, M. G. (2021). Can indigenous and community-based ecotourism serve as a catalyst for land sparing in Latin America? *Journal of Travel Research, 60*(7), 1566–1580.

Ickowitz, A., McMullin, S., Rosenstock, T., Dawson, I., Rowland, D., Powell, B., & Nasi, R. (2022). Transforming food systems with trees and forests. *The Lancet Planetary Health, 6*(7), e632–e639.

Jaramillo, C. (2023). The evolution of extant South American tropical biomes. *New Phytologist, 239*(2), 477–493.

Karki, G., Kunwar, R., Bhatta, B., & Devkota, N. R. (2022). Climate change effects, adaptation and community-based forest management in the mid-hills of Tanahu and Kaski districts, Nepal. *International Forestry Review, 24*(4), 573–593.

Kor, L., Homewood, K., Dawson, T. P., & Diazgranados, M. (2021). Sustainability of wild plant use in the Andean Community of South America. *Ambio, 50*(9), 1681–1697.

Lombe, P., Carvalho, E., & Rosa-Santos, P. (2024). Drought dynamics in sub-Saharan Africa: Impacts and adaptation strategies. *Sustainability, 16*(22), 9902.

Mantey, P. P. K., & Teye, J. K. (2021). Forest dependence among rural households in southern Ghana: Implications for conservation and poverty reduction. *Ghana Journal of Geography, 13*(1), 1–24.

Miller, A. B., Cox, C., & Morse, W. C. (2023). Ecotourism, wildlife conservation, and agriculture in Costa Rica through a social-ecological systems lens. *Frontiers in Sustainable Tourism, 2*, 1179887.

Mugari, E., Nethengwe, N. S., & Gumbo, A. D. (2024). The utilization and contribution of timber and non-timber forest products to livelihoods under a changing climate in the Limpopo River Basin. *Environmental Research Communications, 6*(2), Article 025005.

Muluneh, M. G. (2021). Impact of climate change on biodiversity and food security: A global perspective—A review article. *Agriculture & Food Security, 10*(1), 1–25.

Muthee, K., Duguma, L., Majale, C., Mucheru-Muna, M., Wainaina, P., & Minang, P. (2022). A quantitative appraisal of selected agroforestry studies in the Sub-Saharan Africa. *Heliyon, 8*(9).

Ngavom, Z., Fotso-Nguemo, T. C., Vondou, D. A., Fotso-Kamga, G., Zebaze, S., Yepdo, Z. D., & Diedhiou, A. (2024). Projected changes in population exposure to extreme precipitation events over Central Africa under the global warming levels of 1.5°C and 2°C: Insights from CMIP6 simulations. *Modeling Earth Systems and Environment, 10*(4), 5753–5769.

Noe, C., Mwamfupe, A., Kweka, O., John, R. W., Silvano, P., Namkesa, F. D., & Ponte, S. (2022). Conservation and development in Tanzania: Background, history, and recent developments. In C. Noe, D. Brockington, & S. Ponte (Eds.), *Contested sustainability: The Political ecology of conservation and development in Tanzania* (pp. 33–70).

Palma Valente, C. A. (2022). Indigenous peoples as guardians of nature. The role of International Environmental and Human Rights Law in the preservation of Amazonia. http://hdl.handle.net/10579/21110

Parthasarathy, S., Kalaivanan, R., Sangavi, R., & Lakshmidevi, P. (2025). *Pests and diseases in spices, plantation and tuber crops*. CRC Press.

Phuong, H. T., Tan, N. Q., Nga, P. T. Q., Chau, D. T. M., Chan, N., Win, K., & Swe, K. N. (2024). Climate change, livelihood resilience, and gender: An intersectional analysis of Vietnam's forest-dependent communities. *Environmental Development, 52*, Article 101072.

Poudyal, B. H., Maraseni, T. N., Aryal, S., & Cockfield, G. (2022). An assessment of selective logging policies and practices in Nepal. In *Agriculture, natural resources and food security: Lessons from Nepal* (pp. 357–374). Springer International Publishing.

Pulhin, J. M., Ramirez, M. A. M., Garcia, J. E., Pangilinan, M. J. Q., Evaristo, M. B. S., Catudio, M. L. R. O., & Mariano, B. J. S. (2024). Contextualizing sustainable forest management and social justice in community-based forest management (CBFM) program in the Philippines. *Trees, Forests and People, 16*, Article 100589.

Rawat, Y. S., & Tekleyohannes, A. T. (2021). Sustainable forest management and forest products industry development in Ethiopia. *International Forestry Review, 23*(2), 197–218.

Reyes-García, V., García-Del-Amo, D., Porcuna-Ferrer, A., Schlingmann, A., Abazeri, M., Attoh, E. M., & Torrents-Ticó, M. (2024). Local studies provide a global perspective of the impacts of climate change on indigenous peoples and local communities. *Sustainable Earth Reviews, 7*(1), 1.

Subedi, B., Poudel, A., & Aryal, S. (2023). The impact of climate change on insect pest biology and ecology: Implications for pest management strategies, crop production, and food security. *Journal of Agriculture and Food Research, 14*, Article 100733.

Tebkew, M., Asfaw, Z., & Worku, A. (2024). The role of agroforestry systems for addressing climate change livelihood vulnerability of farmers of Northwestern Ethiopia. *Heliyon, 10*(17).

Tota, W., Maulud, K. N. A., & Noor, N. M. (2024). Impact of community-based adaptation methods in creating resilient communities in Indonesia and Philippines. In *E3S Web of Conferences* (Vol. 599, p. 05003). EDP Sciences.

Vinodhini, S. M., Manibharathi, S., Pavithra, G., & Sakthivel, S. (2023). Agroforestry: Integrating trees into agricultural systems. In *Recent approaches in agriculture* (p. 246). Elite Publishing House.

WHO. (2022). African Traditional Medicine Day 2022: Message of WHO Regional Director for Africa, Dr Matshidiso Moeti. https://www.afro.who.int/regional-director/speeches-messages/african-traditional-medicine-day-2022

Wunderling, N., Staal, A., Sakschewski, B., Hirota, M., Tuinenburg, O. A., Donges, J. F., & Winkelmann, R. (2022). Recurrent droughts increase risk of cascading tipping events by outpacing adaptive capacities in the Amazon rainforest. *Proceedings of the National Academy of Sciences, 119*(32), Article e2120777119.

Yang, Z., Gao, X., Lei, J., Meng, X., & Zhou, N. (2022). Analysis of spatiotemporal changes and driving factors of desertification in the Africa Sahel. *CATENA, 213*, Article 106213.

CHAPTER 3

Climate Resilience in Fishing Communities in the Global South: Adaptive Management Strategies and Policy Frameworks

Abstract Fishing communities in the Global South are among the most vulnerable to climate change. This chapter explores the climate resilience of fishing communities in the Global South, focusing on adaptive management strategies and policy frameworks that aim to enhance their capacity to cope with climate impacts. It highlights the significant challenges posed by ocean acidification, rising sea temperatures, and shifting fish migration patterns, all of which threaten food security and livelihoods in these communities. The chapter emphasizes the role of Traditional Ecological Knowledge (TEK) in fisheries management, showcasing how community-based resource management and conservation efforts can help build adaptive capacity. Additionally, the diversification of livelihoods, through strategies such as aquaculture, tourism, and agro-fisheries, is examined as a crucial means of reducing vulnerability and enhancing economic resilience. The chapter also reviews national and international policy frameworks aimed at fostering fisheries resilience, while identifying gaps in governance, financing, and community involvement that hinder effective climate adaptation. Finally, it underscores the importance of strengthening local governance and fostering community-led initiatives, which are key to ensuring sustainable and adaptive management practices. Overall, the chapter calls for integrated approaches that combine traditional knowledge, diversified livelihoods, and robust policy support

N. P. Awazi, *Building Resilience*, Palgrave Studies in Climate Resilient Societies, https://doi.org/10.1007/978-3-031-93791-0_3

to secure the long-term resilience of fishing communities in the face of climate change.

Keywords Climate change · Fisheries resilience · Traditional ecological knowledge · Livelihood diversification · Policy frameworks · Governance · Community-based management

3.1 Climate Change Impacts on Fisheries and Fishing Communities in the Global South

The Global South, home to many developing countries, relies heavily on natural resources for both economic development and sustenance. Fisheries, in particular, are a cornerstone of food security, livelihoods, and culture in many of these nations. However, the impacts of climate change, particularly ocean acidification, rising sea temperatures, fish migration, and the resulting effects on food security, pose significant threats to these vulnerable communities (Table 3.1). Ocean acidification (OA) is one of the most alarming consequences of climate change. It refers to the process by which the ocean absorbs excess carbon dioxide (CO_2) from the atmosphere, resulting in a lowering of the ocean's pH and a decrease in carbonate ions. These ions are crucial for marine organisms such as corals, mollusks, and certain plankton species, which depend on them to form their calcium carbonate shells and skeletons (Samanth, 2024). In the Global South, where many coastal communities depend on shellfish and coral reef ecosystems for their livelihoods, OA poses a serious threat. As oceans become more acidic, the growth and reproduction of marine species, particularly those that form the base of the food web, are negatively affected. For example, studies have shown that OA can reduce the survival rates of commercially important fish species such as cod and shellfish species like oysters and clams (Baag & Mandal, 2022). This has serious economic implications for communities in coastal regions, particularly in developing nations where fisheries form a major part of the economy. In countries such as the Philippines, Indonesia, and Senegal, the degradation of reef ecosystems due to acidification further jeopardizes the livelihoods of fishers who depend on coral reef fish species for both commercial sale and subsistence consumption (Ebarvia, 2022; Miao & Nduneseokwu, 2025). Moreover, the impact on marine biodiversity is not limited to the direct effects of OA but extends to disruptions in the entire marine food web. As plankton populations decline due to acidification, the entire ecosystem is altered, affecting species higher up the food chain, including fish. For communities in the Global South that rely on fish for protein, these disruptions can exacerbate malnutrition and poverty.

The warming of the world's oceans is another direct consequence of climate change that disproportionately affects fisheries in the Global South. Sea surface temperatures (SSTs) have risen significantly in recent decades due to increasing global temperatures, with the tropical oceans being particularly vulnerable. Rising SSTs disrupt marine ecosystems in multiple ways. Fish species that are sensitive to temperature shifts, such as tuna and grouper, may experience reduced reproductive success or increased mortality rates as water temperatures rise (Bo et al., 2023; Zhang et al., 2025). For fishing communities, the impacts of rising SSTs go beyond the direct effects on fish populations. Increased water temperatures lead to coral bleaching events, where symbiotic relationships between corals and algae break down due to stress from the heat. The resulting degradation of coral reefs reduces habitat complexity and fish biodiversity (Sobha et al., 2023). Coral reefs are vital to the livelihoods of fishing communities in the Global South, as they provide shelter and food for numerous fish species that are essential for both local consumption and export. Furthermore, warmer waters tend to alter the distribution of fish stocks, often moving them away from traditional fishing grounds. For example, a study by Kim et al. (2023) shows that the warming of oceans will cause fish stocks to shift toward the poles, which may render traditional fishing zones in tropical regions less productive. This shift could increase competition for fish in already crowded regions, disrupting local economies and potentially leading to conflict over resources.

In addition to ocean acidification and rising sea temperatures, climate change also influences fish migration patterns. Many fish species, particularly those in the Global South, exhibit migratory behaviors that are sensitive to environmental changes (Alò et al., 2021; Nagelkerken et al., 2023). These migrations are often linked to the availability of prey, water temperature, and breeding conditions. With climate change altering sea temperatures and ocean currents, the distribution of fish stocks has been undergoing significant changes. Fish populations are increasingly migrating to new areas in search of more suitable habitats. This phenomenon, known as "species range shifts," is already being observed in regions like the eastern Indian Ocean, where the distribution of tuna

Table 3.1 Impacts of climate change on fisheries and fishing communities in the Global South

Impact	*Description*	*Examples*	*Key references*
Ocean Acidification (OA)	OA results from increased CO_2 in the atmosphere, lowering ocean pH and affecting marine organisms, particularly shellfish and corals	OA negatively impacts shellfish populations in coastal regions such as the Philippines, Indonesia, and Senegal	Samanth (2024), Baag and Mandal (2022), Ebarvia (2022), Miao and Nduneseokwu (2025)
Rising sea temperatures (SSTs)	Rising SSTs disrupt marine ecosystems, affecting fish reproduction, coral reefs, and the distribution of fish stocks	The warming of oceans affects tuna and grouper populations, leading to coral bleaching and changes in fish distribution	Bo et al. (2023), Sobha et al. (2023), Kim et al. (2023), Zhang et al. (2025)
Fish migration and range shifts	Changes in sea temperature and currents alter fish migration patterns, causing species to move to new areas in search of suitable habitats	Tuna and other fish species in the eastern Indian Ocean shifting their distribution, impacting traditional fishing grounds	Alò et al. (2021), Nagelkerken et al. (2023), Dimarchopoulou et al. (2022), Neokye and Wang (2024), Briscoe et al. (2025)
Food security	Declining fish populations affect food security, especially in countries where fish provide a major protein source	Coastal populations in Mozambique, Ecuador, and Bangladesh rely on fish as a major protein source, which is threatened by climate change impacts	Bernzen et al. (2022), El-Sayed and Fitzsimmons (2023), Arthur et al. (2022)

(continued)

Table 3.1 (continued)

Impact	*Description*	*Examples*	*Key references*
Economic impacts	Decline in fish stocks reduces income from both local consumption and exports, leading to economic hardship for fishing communities	Loss of fish stocks affects the livelihoods of small-scale fishers and national economies in Global South countries	Paniagua and Rayamajhee (2024), Mielgo et al. (2024)
Marine biodiversity disruption	Ocean acidification and warming disrupt marine food webs, affecting biodiversity and the availability of key fish species	Coral reef degradation and declines in plankton populations disrupt entire marine ecosystems, reducing fish availability	Samanth (2024), Baag and Mandal (2022), Sobha et al. (2023)

and other commercially important fish is changing (Briscoe et al., 2025; Dimarchopoulou et al., 2022; Neokye & Wang, 2024). These migrations not only affect the sustainability of fisheries in their traditional zones but also raise the possibility of overfishing in new areas that may be ill-equipped to manage these increased pressures. For fishing communities in the Global South, this shift is particularly challenging. Many of these communities are small-scale, artisanal fishers who rely on established fishing territories. When fish stocks move beyond these zones, these fishers may find themselves without adequate catches, leading to a decline in income and food security. Furthermore, these communities often lack the resources or technology to adapt to these new migration patterns, such as developing new fishing techniques or relocating to different regions. The impacts of fish migration are particularly acute in the context of regional and international fisheries management. Many Global South countries depend on cooperative management arrangements to sustainably harvest marine resources. However, as fish stocks move across national boundaries due to climate change, these management systems are increasingly strained. A lack of coordination between countries can lead to overfishing, as nations may seek to exploit these resources before others can (Mielgo et al., 2024; Paniagua & Rayamajhee, 2024).

At its core, the disruption of fisheries due to climate change directly impacts food security in the Global South. Fisheries in many of these regions provide not only a source of income but also an essential source of nutrition. Fishes are a major protein source for coastal populations, especially in countries like Mozambique, Ecuador, and Bangladesh, where fish contribute up to 50% of animal protein intake (Bernzen et al., 2022; El-Sayed & Fitzsimmons, 2023). As fish populations become less predictable and available due to climate change, the nutritional security of these communities is at risk. Small-scale fishers, who are often the most vulnerable, may face diminished catches, resulting in reduced access to food and income. In the face of declining fisheries, many may resort to coping strategies such as increased reliance on cheaper and less nutritious food, which can exacerbate malnutrition. The relationship between fisheries and food security is further complicated by the economic dependence of many fishing communities on the global market. Fisheries in the Global South are not only crucial for local consumption but also represent significant export earnings. The loss of these fisheries or a decline in fish stocks can lead to economic hardship and a reduction in the overall availability of fish both locally and internationally. A reduction in fish stocks could also lead to increased competition for remaining fish, raising prices and making fish less accessible to vulnerable populations (Arthur et al., 2022).

3.2 Traditional Ecological Knowledge and Adaptive Management in Fishing Communities in the Global South

Traditional Ecological Knowledge (TEK) refers to the cumulative and evolving body of knowledge, practices, and beliefs held by indigenous and local communities regarding the relationships between living beings, including humans, and their environment (Sinthumule, 2023). In the context of fishing communities, TEK plays a pivotal role in managing fisheries, conserving biodiversity, and sustaining local livelihoods. This body of knowledge is not static but is adaptive, evolving with the ecological, social, and economic changes over time. In the Global South, where fishing communities often rely on local natural resources for their subsistence, TEK can be a crucial component of sustainable resource management and conservation.

In the Global South, where many communities rely heavily on small-scale fisheries, TEK represents a deep, context-specific understanding of local ecosystems and resource dynamics. This knowledge is often passed down through generations and is deeply intertwined with the social and cultural fabric of fishing communities. One example is the knowledge held by the coastal communities in the Philippines. A study by Kotowicz et al. (2022) and Obiero et al. (2023) highlights the role of traditional fishing practices and the associated knowledge in managing local marine resources. In this context, fishing communities use traditional community-based seasonal fishing techniques, which regulates when and where fish can be harvested, ensuring that fish stocks are not overexploited. This practice, underpinned by TEK, helps maintain sustainable fish populations by aligning fishing activities with ecological rhythms. Similarly, the indigenous fishing communities of the Amazon, such as the indigenous Kayapo people, possess TEK that informs their management of riverine fish stocks. These communities are known to use seasonal and spatial closures of fishing zones, which helps fish populations recover. According to Bagsit et al. (2021), such traditional management practices are key to ensuring long-term sustainability in these communities. These traditional practices are not only effective in regulating harvests but also in maintaining biodiversity and ecosystem health.

Community-Based Resource Management (CBRM) is an approach that emphasizes local community participation in the management of natural resources, often relying on TEK as a foundation. CBRM is especially important in the Global South, where many communities have a direct connection to the resources they manage, and where external, top-down management approaches have often failed to achieve sustainable results. A prominent case of CBRM in the Global South is the management of coral reefs in Fiji. According to studies by Kitolelei et al. (2021) and Riera (2022), traditional tabu (or no-take) areas established by Fijian fishing communities have been instrumental in replenishing fish stocks. These tabus are a direct expression of TEK, reflecting a deep understanding of when and where fish populations are vulnerable and how human activities impact marine ecosystems. These areas have been shown to have higher fish biomass compared to unprotected areas, demonstrating the effectiveness of community-based management underpinned by TEK. Other examples come from the coastal communities of Madagascar and Tanzania. Here, the management of sea cucumber fisheries has been traditionally overseen by local communities using TEK that dictates fishing seasons and areas. Research by Chengula (2023) and Oloko et al. (2024) emphasizes the importance of integrating TEK with formal fisheries regulations to avoid overexploitation. In Madagascar and Tanzania, local knowledge of seasonal changes in sea cucumber populations and the environmental factors that influence them has been used to establish community-led monitoring systems, ensuring that resource extraction does not exceed sustainable levels. These examples illustrate how TEK informs CBRM efforts by providing localized knowledge of ecosystems, which enables better-targeted interventions to sustain fisheries and conserve biodiversity. TEK also fosters community cohesion and collective responsibility, which are critical for the long-term success of resource management efforts.

In the Global South, conservation efforts increasingly recognize the importance of integrating TEK into the management of protected areas and conservation landscapes. Indigenous and local knowledge systems often embody centuries of experience in observing and managing ecosystems, and they offer valuable insights for contemporary conservation challenges. A key example of conservation efforts informed by TEK is the community-led conservation initiative in the Mesoamerican Barrier Reef System (MBRS) along the coasts of Belize, Honduras, and Guatemala. According to research by Cinti et al. (2025), local fishing communities in this region have historically implemented a range of conservation measures based on TEK, such as seasonal closures, selective harvesting methods, and the establishment of marine protected areas (MPAs). These practices are grounded in local knowledge about the breeding cycles of key species and the environmental factors that influence the health of coral reef ecosystems. Over the years, local community-based approaches have proven to be more successful than top-down, government-led conservation measures in sustaining marine biodiversity and fish stocks. In India, the Orissa Marine Resources Conservation Consortium (OMRCC) integrates TEK into conservation efforts along the eastern coast. The community of Chilika Lake, known for its rich biodiversity, has used TEK to create management systems that balance conservation and sustainable use. The lake is one of the largest brackish water lagoons in the world and an important breeding ground for several fish species. According to studies by Singhar and Gundimeda (2024), the local communities in the area have long understood the importance of seasonal fishing bans during spawning periods. These traditional practices have played an essential role in maintaining the health of fish populations in the lagoon and promoting sustainable fishery yields. These examples demonstrate the capacity of traditional knowledge to contribute to modern conservation efforts. When combined with scientific knowledge, TEK offers a more holistic and context-sensitive approach to conservation, one that incorporates both ecological and cultural dimensions of biodiversity management.

In fishing communities of the Global South, where ecosystems are often dynamic and unpredictable, adaptive management can be particularly valuable. TEK, with its inherent capacity to respond to changing conditions, aligns well with adaptive management principles. A prime example of adaptive management in a fishing context is the management of the mangrove ecosystems in coastal Bangladesh. Research by Mahmood et al. (2021) shows how communities along the Sundarbans have historically adjusted their fishing practices based on their observations of mangrove health and fish migration patterns. The adaptive strategies of these communities include adjusting fishing techniques based on changes in salinity, water levels, and the presence of key species. These adaptations demonstrate the value of TEK as a form of adaptive management, where learning from the environment guides resource use decisions. Furthermore, in the Gulf of Mexico, communities have implemented adaptive management strategies based on traditional knowledge regarding the lifecycle of the totoaba fish, a species at risk of extinction due to overfishing. According to Morales-Bojórquez et al. (2021), local fishers have long known the seasonal patterns of totoaba and have adjusted their fishing efforts accordingly. Adaptive management, based on both scientific and traditional knowledge, has helped implement effective conservation measures, including seasonal fishing restrictions and fishing gear modifications, aimed at preventing overfishing and promoting sustainable harvesting practices. In these cases, adaptive management provides a framework for integrating TEK into decision-making, ensuring that resource management is flexible and responsive to environmental changes and new information.

Thus, Traditional Ecological Knowledge (TEK) plays an indispensable role in fisheries management, community-based resource management, and conservation efforts in the Global South. TEK offers a deep, context-specific understanding of local ecosystems and resource dynamics, which can be integrated into contemporary approaches such as adaptive management to enhance the sustainability of fisheries and conservation practices. Case studies from regions like the Philippines, Tanzania, Fiji, Madagascar, and the Sundarbans demonstrate the effectiveness of TEK in managing natural resources, conserving biodiversity, and sustaining local livelihoods. As climate change and global environmental challenges continue to impact ecosystems, the integration of TEK with modern management practices offers a promising pathway for ensuring the resilience of fishing communities and their ecosystems. By combining the strengths of traditional knowledge with adaptive management approaches, fishing communities in the Global South can build more sustainable and equitable resource management systems that benefit both people and the environment.

3.3 Diversification of Livelihoods in Fishing Communities in the Global South

Fishing communities in the Global South have traditionally relied on fishing as their primary source of livelihood, with their well-being and economic stability deeply tied to marine and freshwater ecosystems. However, over the years, various challenges have emerged, including environmental degradation, overfishing, climate change, and global market fluctuations, all of which have threatened the sustainability of fishing-based livelihoods. In response to these challenges, many fishing communities have embraced diversification of livelihoods as a strategy to reduce vulnerability and ensure economic resilience. Diversification involves expanding economic activities beyond traditional fisheries to include alternative livelihoods, such as aquaculture, tourism, agro-fisheries, and other complementary sectors (Table 3.2).

The dependence of fishing communities on natural resources, particularly fisheries, makes them highly susceptible to environmental and economic shocks. In many parts of the Global South, fishing resources are being overexploited, leading to declining fish stocks and diminished catches. Climate change is exacerbating this vulnerability through rising sea temperatures, acidification, and extreme weather events such as cyclones and droughts (Jha & Dev, 2024). For instance, in the coastal regions of Sub-Saharan Africa, the livelihoods of artisanal fishermen are threatened by declining fish catches and changes in oceanic conditions (Muringai et al., 2021). Diversification of livelihoods, therefore, provides an opportunity for fishing communities to reduce their dependence on a single resource and hedge against the risks posed by environmental changes. It also offers new avenues for income generation, improves food

Table 3.2 Diversification of livelihoods in fishing communities in the Global South

Livelihood diversification strategy	*Description*	*Examples*	*Key references*
Aquaculture (Fish farming)	Fish farming provides a controlled environment for cultivating fish and other aquatic species, offering a steady supply of seafood and reducing pressure on wild fish stocks	Shrimp farming in coastal regions of Bangladesh has created jobs, especially for women, and contributed to economic growth	Boyd et al. (2022), Shanta et al. (2025), Ho et al. (2025)
Tourism (ecotourism and recreational fishing)	Coastal and marine tourism has become a significant source of income through ecotourism, recreational fishing, and cultural experiences	Fishing communities in Palawan and Bohol, Philippines, have shifted to ecotourism, offering coral reef exploration and traditional fishing practices to tourists	Fischer et al. (2021), Brillo and Simondac-Peria (2021), Leotaud et al. (2024), Bennett et al. (2024)
Agro-fisheries (rice and fish farming)	Integration of agriculture (such as rice farming) and aquaculture, allowing for diversified income from both aquatic and terrestrial production	The Mekong Delta in Vietnam uses rice-fish farming, where fish are raised in flooded rice fields alongside crop cultivation	Sainjoo (2024)
Non-Timber Forest Products (NTFPs)	Harvesting and selling products like honey, medicinal plants, and forest products offer additional income sources for fishing communities in rural areas	In Uganda and Tanzania, smallholder farmers integrate fish farming with crop cultivation while harvesting NTFPs for income	Roscher et al. (2022), March and Failler (2022)

(continued)

Table 3.2 (continued)

Livelihood diversification strategy	*Description*	*Examples*	*Key references*
Alternative energy (solar power)	Adoption of renewable energy sources, such as solar energy, to generate income, improve energy access, and support other activities like irrigation	Solar-powered irrigation systems are being introduced to support agriculture in areas lacking electricity	Caballero et al. (2023)
Challenges in diversification	Diversification can increase community resilience but faces challenges like access to resources, skills, capital, and markets, especially for new activities like aquaculture and tourism	Over-reliance on a single diversification strategy like aquaculture can expose communities to risks such as disease outbreaks or environmental degradation	Agyei (2022)

security, and strengthens community resilience. However, the success of diversification strategies depends on various factors, including local context, access to resources, institutional support, and the capacity of community members to adapt to new economic activities.

Aquaculture, or fish farming, has emerged as a prominent alternative livelihood strategy for fishing communities in the Global South. Aquaculture offers a controlled environment for the cultivation of fish and other aquatic species, enabling communities to produce a steady supply of seafood without relying solely on wild fish stocks. According to Boyd et al. (2022), aquaculture has the potential to alleviate pressure on natural fisheries by providing an alternative source of protein and income for coastal populations. In countries such as Bangladesh, India, and Vietnam, aquaculture has become a significant livelihood activity. For instance, in Bangladesh, the development of shrimp farming in coastal regions has created numerous jobs for local communities, particularly women, and contributed to the country's economic growth (Shanta et al., 2025). However, the expansion of aquaculture is not without challenges. Environmental concerns such as the destruction of mangrove ecosystems, pollution from farmed fish waste, and disease outbreaks are significant issues that need to be managed carefully (Ho et al., 2025). Therefore, aquaculture must be integrated into a broader framework of sustainable resource management to ensure long-term viability.

Tourism is another diversification strategy that many fishing communities in the Global South have adopted. Coastal and marine tourism, including ecotourism and recreational fishing, has become a significant source of income for communities, especially those in coastal regions with rich natural and cultural heritage. Tourists flock to tropical beaches, coral reefs, and fishing villages, creating opportunities for local entrepreneurs to engage in hospitality, boat tours, and cultural experiences. In the Philippines, for example, fishing communities in regions like Palawan and Bohol have diversified into ecotourism, offering visitors the chance to explore coral reefs, snorkel, and experience traditional fishing practices. These tourism ventures not only generate income for local households but also raise awareness about the importance of marine conservation (Brillo & Fischer et al., 2021; Simondac-Peria, 2021). Similarly, in the Caribbean, the development of sustainable tourism in places like the Bahamas and Belize has allowed fishing communities to tap into global tourism markets while preserving their marine environments (Bennett et al., 2024; Leotaud et al., 2024). However, tourism can be a double-edged sword. If not carefully managed, tourism development can lead to environmental degradation, such as coral reef damage from over-tourism, pollution, and the displacement of local communities. Therefore, sustainable tourism practices, which emphasize community involvement, conservation, and equitable distribution of benefits, are crucial for ensuring that tourism serves as a viable and long-term alternative livelihood.

Agro-fisheries, which involves the integration of agriculture and fisheries, is another diversification strategy that holds promise for fishing communities in the Global South. This approach involves the cultivation of crops, such as rice, in close proximity to fish farming activities, creating synergies between agriculture and aquaculture. Agro-fisheries systems, also known as integrated farming systems (IFS), can help reduce dependency on fishing alone by creating diverse income streams from both aquatic and terrestrial production. For example, in the Mekong Delta of Vietnam, many small-scale farmers and fishermen have adopted rice-fish farming systems. This system allows farmers to cultivate rice while simultaneously raising fish in flooded rice fields. Not only does this approach provide additional income, but it also contributes to increased biodiversity and better soil quality. Furthermore, agro-fisheries have been shown to enhance food security by ensuring a more diverse and reliable food supply (Sainjoo, 2024). In some parts of Africa, agro-fisheries are also being promoted as a means of diversifying livelihoods in communities where agriculture is the dominant activity. For instance, in Uganda and Tanzania, smallholder farmers are encouraged to integrate fish farming with crop cultivation. This diversification strategy helps increase household income and mitigate the risks posed by fluctuations in agricultural production due to climate variability.

Fishing communities in the Global South are increasingly turning to alternative energy sources and the exploitation of non-timber forest products (NTFPs) as additional livelihood activities. In some coastal areas, local communities have adopted the production of renewable energy sources, such as solar energy, to generate income and improve energy access (Caballero et al., 2023). Solar-powered irrigation systems, for example, are being introduced to support agriculture in areas where access to electricity is limited. Similarly, the collection and sale of NTFPs, such as honey, medicinal plants, and forest products, provide alternative income sources for fishing households in rural areas. These products are often harvested from local ecosystems and can be sold in local markets or even exported. Through diversification into such sectors, fishing communities reduce their vulnerability to changes in fish stocks and create more sustainable livelihoods (March & Failler, 2022; Roscher et al., 2022).

The diversification of livelihoods in fishing communities can have significant social, economic, and environmental implications. On the positive side, diversification helps to reduce poverty by creating new income sources, enhances food security, and strengthens community resilience against environmental and economic shocks. It also provides an opportunity for women and marginalized groups to participate in economic activities that were previously inaccessible to them, thus contributing to gender equality and social inclusion (Roscher et al., 2022). However, the success of diversification strategies depends on the availability of resources, appropriate skills, and supportive policies. For example, while aquaculture and tourism offer promising economic opportunities, they require access to capital, training, and market access. Moreover, over-reliance on a single diversification strategy, such as aquaculture, can also expose communities to new risks, such as disease outbreaks or environmental degradation (Agyei, 2022). A diversified approach that incorporates a range of complementary livelihood activities is thus essential for achieving long-term sustainability.

3.4 Policy Frameworks and Gaps for Climate Resilience in Fishing Communities in the Global South

Fishing communities in the Global South are increasingly facing the challenges posed by climate change, such as rising sea levels, extreme weather events, and shifts in marine ecosystems. Climate change affects the livelihoods of millions who rely on fishing, leading to food security risks, economic instability, and social distress. To address these vulnerabilities, various national and international policy frameworks have been developed to support resilience in these communities, although there remain significant gaps in governance, financing, and community involvement, limiting the effectiveness of these policies (Table 3.3).

At the national level, many countries in the Global South have adopted climate change adaptation policies that integrate fisheries resilience (de Carvalho et al., 2023). These policies often aim to mitigate the impact of climate change on fish stocks, protect vulnerable fishing communities, and promote sustainable fisheries management. However, there is considerable variation in how these policies are structured and implemented. The most common national climate resilience policies for the fisheries sector are climate change adaptation strategies, coastal and marine protection, as well as fisheries management and sustainability (Bryndum-Buchholz et al., 2021). Several countries, particularly those in Sub-Saharan Africa, Southeast Asia, and Latin America, have developed national climate change adaptation strategies that include measures to protect fisheries. For example, the African Union's "Climate Change and Resilience Framework" emphasizes the importance of climate-smart fisheries management, ecosystem-based adaptation, and enhancing the adaptive capacity of coastal communities (AU Climate Change Strategy, 2022). Similarly, the Philippines has integrated climate change adaptation into its national fisheries policies through its "Philippine Fisheries Code," which addresses resource sustainability and disaster resilience (Manlosa et al., 2021; Philippines Fisheries Code, 1998). National governments

Table 3.3 Policy frameworks and gaps for climate resilience in fishing communities in the Global South

Policy Framework	*Description*	*Key examples*	*Key references*
National Climate Adaptation Policies	National policies aimed at integrating fisheries resilience, focusing on climate change adaptation, sustainable fisheries management, and coastal protection	• African Union's "Climate Change and Resilience Framework": Promotes climate-smart fisheries management and ecosystem-based adaptation (AU Climate Change Strategy, 2022) • Philippine Fisheries Code: Focuses on resource sustainability and disaster resilience (Philippine Fisheries Code, 1998)	de Carvalho et al. (2023), Manlosa et al. (2021), Bryndum-Buchholz et al. (2021)
Policies for Marine Protected Areas (MPAs)	Policies promoting the establishment of MPAs to protect vulnerable marine ecosystems and enhance biodiversity, supporting resilience to climate impacts	Countries like Kenya and Tanzania have established MPAs to protect marine ecosystems. Costa Rica's National Coastal Zone Management Program integrates climate change adaptation for sustainable fisheries (Sorensen, 1990)	Kirkman et al., (2023), Arafeh-Dalmau et al. (2023)
International frameworks and conventions	Global frameworks for transboundary cooperation on fisheries management and climate resilience, including agreements on migratory fish stocks, illegal fishing, and climate adaptation	• UNFCCC and Paris Agreement: Encourage climate adaptation plans (NAPs) but focus more on mitigation • FAO Voluntary Guidelines for Securing Sustainable Small-Scale Fisheries: Focuses on community participation and ecosystem-based management (FAO Guidelines) • Regional Fisheries Management Organizations (RFMOs): For example, the Indian Ocean Tuna Commission (IOTC)	Nakamura (2022), Coates et al. (2023), Łukaszuk (2024), Islam and Hasan (2024), Matovu (2024)

(continued)

Table 3.3 (continued)

Policy Framework	*Description*	*Key examples*	*Key references*
Sustainable Development Goals (SDGs)	Global development framework with a focus on marine conservation, with SDG 14 specifically dedicated to ocean and marine resource sustainability, including resilience for fishing communities	SDG 14 encourages the conservation of oceans and resilience of coastal communities, but many Global South countries face challenges in fully integrating these goals (SDG 14)	Evans et al. (2023)
Regional cooperation agreements	Policies that encourage regional cooperation to manage shared fish stocks and marine resources under climate change pressures	IOTC and SEAFO (South East Atlantic Fisheries Organization) promote regional cooperation for sustainable fisheries management (IOTC)	Łukaszuk (2024), Evans et al. (2023)
Governance gaps	Weak, fragmented governance structures hinder effective management of fisheries and climate resilience. Issues include inadequate monitoring, weak enforcement, and political instability	• Lack of integrated fisheries policies in broader climate adaptation strategies • Fragmented governance across different levels of government	Andrews et al. (2021), Umamaheswari et al. (2021)
Financing gaps	Insufficient financial resources and slow delivery from international mechanisms hinder fisheries resilience efforts. Limited involvement of the private sector in financing resilience initiatives	• Green Climate Fund: Slow to deliver resources for fisheries resilience • Lack of innovative financial mechanisms to support sustainable fisheries (e.g., financing from private sector)	Andrews et al. (2021), Evans et al. (2023)
Community involvement gaps	Lack of meaningful community participation in policy development and implementation, excluding marginalized groups, which reduces policy effectiveness	• Exclusion of small-scale fishers, women, and indigenous peoples from decision-making processes • Insufficient capacity-building for local stakeholders to manage climate change impacts on fisheries	Evans et al. (2023), Umamaheswari et al. (2021)

in the Global South are increasingly focusing on marine protected areas (MPAs) as a tool for enhancing climate resilience (Arafeh-Dalmau et al., 2023; Kirkman et al., 2023). MPAs can serve as refuges for fish species under stress from climate change, helping to maintain biodiversity and restore fish stocks. Countries like Kenya and Tanzania have established MPAs to protect vulnerable marine ecosystems and fisheries. In Latin America, Costa Rica's National Coastal Zone Management Program has integrated climate change adaptation into coastal zone governance, with a particular focus on sustainable fisheries (Sorensen, 1990). National fisheries policies often aim to balance the need for economic development with sustainable resource use. In countries such as Indonesia, Brazil, and India, there is a growing recognition of the importance of maintaining healthy fish stocks for both food security and economic stability. However, the actual implementation of sustainable fisheries management practices often faces challenges, including overfishing, illegal, unreported, and unregulated (IUU) fishing, and weak enforcement mechanisms (Widjaja et al., 2023).

International frameworks (policies, conventions and agreements), play a crucial role in supporting fisheries resilience in the Global South (Coates et al., 2023; Islam & Hasan, 2024; Łukaszuk, 2024; Matovu, 2024; Nakamura, 2022). These frameworks set global standards and encourage cooperation between countries to address transboundary issues such as migratory fish stocks, illegal fishing, and climate-induced impacts on fisheries. The most prominent of these international policies, agreements and conventions for fisheries resilience in the Global South include: the United Nations Framework Convention on Climate Change (UNFCCC); the Food and Agriculture Organization (FAO) and its Voluntary Guidelines for Securing Sustainable Small-Scale Fisheries; the High Seas Treaty; the United Nations Sustainable Development Goals (SDGs) as well as Regional Cooperation Agreements. Under the UNFCCC, countries are encouraged to develop national adaptation plans (NAPs) to address the impacts of climate change, including on fisheries. The Paris Agreement, adopted in 2015, includes a global goal to limit global warming to well below 2 °C, which indirectly impacts the resilience of marine ecosystems. However, the focus of the UNFCCC on greenhouse gas mitigation, rather than adaptation, has limited attention to fisheries resilience specifically. The FAO has developed frameworks such as the "Voluntary Guidelines for Securing Sustainable Small-Scale Fisheries," which emphasize the need for community participation, gender equity, and ecosystem-based approaches to fisheries management. These guidelines have been widely endorsed by countries in the Global South but are non-binding, which limits their implementation in the absence of strong national policies. The SDGs provide a comprehensive framework for global development, including Goal 14, which focuses on the conservation and sustainable use of oceans, seas, and marine resources. SDG 14 encourages countries to enhance the resilience of coastal communities and marine ecosystems, but many Global South countries struggle to fully integrate these goals into national policies due to resource constraints and competing priorities. Many countries in the Global South are part of regional fisheries management organizations (RFMOs) that aim to manage shared fish stocks and marine resources. For example, the Indian Ocean Tuna Commission (IOTC) and the South East Atlantic Fisheries Organization (SEAFO) promote regional cooperation in managing fish stocks in the face of climate change. These organizations have contributed to improved governance of shared marine resources, though enforcement remains a challenge.

Despite these policies and frameworks, several gaps remain in addressing the climate resilience of fishing communities in the Global South. These gaps fall into three major categories which are governance, financing, and community involvement (Andrews et al., 2021; Evans et al., 2023; Umamaheswari et al., 2021). Governance structures in many countries in the Global South are often weak, fragmented, and unable to effectively manage fisheries in the face of climate change. Inadequate monitoring and enforcement of fishing regulations, lack of coordination between different levels of government, and poor implementation of international commitments undermine the resilience of fishing communities. In many cases, fisheries policies are not integrated into broader climate change adaptation strategies, leading to fragmented and ineffective responses. Moreover, political instability and corruption further hinder the implementation of sound fisheries governance. Adequate financing is crucial for building climate resilience in fishing communities, but there is a significant gap in the availability of financial resources for adaptation measures. Many countries in the Global South face budget constraints and are unable to allocate sufficient funds to support the implementation of climate resilience policies. International financing mechanisms, such as the Green Climate Fund, have been slow to deliver resources to support fisheries resilience, and the funds that are available are often difficult to access due to bureaucratic barriers. Additionally, the private sector's involvement in financing fisheries resilience is limited, and there is a lack of innovative financial mechanisms that could support the transition to sustainable fisheries. Effective climate resilience strategies must be grounded in the knowledge, needs, and priorities of local fishing communities. However, there is often a lack of meaningful community participation in policy development and implementation. In many countries, top-down decision-making processes exclude local stakeholders, particularly marginalized groups such as women, indigenous peoples, and small-scale fishers. This lack of inclusion results in policies that do not reflect the realities on the ground, reducing their effectiveness in enhancing resilience. Moreover, there is often insufficient capacity-building for local communities to engage in climate adaptation efforts and manage the impacts of climate change on fisheries.

3.5 Strengthening Local Governance and Participation for Climate Resilience in Fishing Communities in the Global South

In the context of climate change, fishing communities in the Global South are particularly vulnerable. These communities are often located in coastal or rural areas that are highly exposed to the effects of rising sea levels, ocean acidification, changing fish stocks, and extreme weather events. Strengthening local governance and enhancing community-led initiatives are pivotal in building resilience and adaptive capacities in these communities.

3.5.1 Understanding Local Governance in the Context of Climate Resilience

For fishing communities, local governance plays a critical role in organizing and coordinating responses to climate impacts and ensuring that these responses are tailored to the specific needs and realities of the community. In many parts of the Global South, fishing communities are often marginalized, with limited access to decision-making processes at higher levels of governance. The role of local governance becomes even more significant in these contexts because it provides the mechanism through which communities can negotiate, advocate, and collaborate on climate adaptation strategies (Gisevius et al., 2025). Local governance ensures that climate resilience strategies reflect the lived experiences and knowledge of the community, which is essential for the effectiveness and sustainability of any interventions.

3.5.2 *The Importance of Community-Led Initiatives*

When local fishing communities in the Global South are given the autonomy to lead in climate adaptation and resilience-building, they are more likely to develop innovative, context-specific solutions that are both effective and sustainable (Dias et al., 2023). One key aspect of community-led initiatives is the integration of traditional knowledge systems and practices. Indigenous and local knowledge, passed down through generations, can offer valuable insights into sustainable resource management and adaptive strategies in the face of environmental change. For example, fishing communities often have intricate knowledge of seasonal changes, local weather patterns, and migratory fish behaviors, which can inform climate adaptation strategies (Okeke-Ogbuafor et al., 2022). Empowering these communities to use and adapt this knowledge in concert with modern scientific approaches can enhance resilience to climate change. Moreover, community-led initiatives foster a sense of ownership and responsibility, which is critical for long-term sustainability. When community members actively participate in decision-making and implementation processes, they are more likely to continue engaging with resilience strategies and ensure that these strategies are maintained over time. This contrasts with top-down approaches, which often lack the necessary local support and fail to account for the specific needs and capacities of the community.

3.5.3 *Local Governance and Climate Resilience*

While local governance plays a vital role in strengthening climate resilience, there are several challenges that fishing communities in the Global South face in enhancing their governance capacities. One significant challenge is the lack of financial resources. Many local governments in the Global South operate with limited budgets, which can restrict their ability to implement comprehensive climate adaptation strategies (Coger et al., 2022; Omukuti et al., 2022). Additionally, local governments may lack the technical expertise or capacity to address climate change issues effectively. This is particularly true in small-scale fishing communities, where local governance structures may be underdeveloped and struggle to support resilience-building efforts. Furthermore, political instability and weak institutions can undermine the effectiveness of local governance. In some countries, local governments are often poorly equipped to address environmental issues due to corruption, inadequate infrastructure, or political interference. This weak governance environment can make it difficult for fishing communities to advocate for their needs or gain access to external support. However, despite these challenges, there are several opportunities to strengthen local governance for climate resilience. One of the most promising opportunities is the rise of decentralized governance models, which aim to devolve decision-making powers to local levels. This approach enables local governments and communities to have greater control over the decisions that affect their lives. For instance, the Philippines has implemented a decentralized governance model for coastal resource management that empowers local governments and community-based organizations to manage their coastal and marine resources (Andriesse et al., 2022; Boncan, 2021; Jack-Kadioglu, 2022). Such models enable local communities to play a more direct role in climate adaptation and resilience-building. Moreover, local governance can be strengthened through multi-level governance arrangements that foster collaboration between local, regional, and national authorities, as well as non-governmental organizations (NGOs) and community-based organizations (CBOs). This approach can facilitate the sharing of resources, expertise, and best practices, thereby enhancing the capacity of local governments and communities to implement effective climate adaptation strategies.

3.5.4 The Role of Local Knowledge and Capacity-Building

Capacity-building is a critical aspect of strengthening local governance and participation in climate resilience. In fishing communities, capacity-building involves enhancing both the technical and organizational skills of local actors, including community members, local government officials, and civil society organizations (Okyere et al., 2023; Simon & Muñoz, 2025). This includes training in areas such as disaster risk management, sustainable resource management, and climate adaptation planning. Local knowledge, as previously mentioned, plays a pivotal role in climate resilience. The integration of local knowledge with scientific data is essential for developing climate adaptation strategies that are culturally relevant and effective. The incorporation of local knowledge systems into formal governance processes can create synergies that improve the design and implementation of climate resilience initiatives. For example, in coastal fishing communities, local knowledge about tidal patterns, fish migration, and storm surge predictions can be used in conjunction with scientific models to design more accurate climate forecasts and early warning systems.

3.5.5 *Case Studies of Community-Led Initiatives in the Global South*

Several case studies highlight the success of community-led initiatives in building climate resilience in fishing communities in the Global South. In the Maldives, for example, local communities have implemented coastal protection measures such as mangrove restoration and coral reef rehabilitation. These initiatives, which are driven by local communities in collaboration with local governments and NGOs, have helped to protect coastal areas from storm surges and rising sea levels (Hilmi et al., 2023; Jauharee, 2022; Nakamura et al., 2023). In another example, the coastal fishing communities in Kenya have been engaged in community-based marine resource management (CBMRM). This approach empowers local fishing communities to manage and conserve their marine resources in a sustainable manner, thereby improving resilience to the impacts of overfishing and climate change (Kamau et al., 2021; Nyawade et al., 2021; Onyango et al., 2021). The success of CBMRM in Kenya highlights the importance of local knowledge, community engagement, and collaborative governance in building resilience.

References

Agyei, B. P. (2022). *Sustainability of pond and cage fish farming systems in the Ashanti, Bono, Eastern and Volta regions in Ghana* (Master's thesis, Norwegian University of Life Sciences).

Alò, D., Lacy, S. N., Castillo, A., Samaniego, H. A., & Marquet, P. A. (2021). The macroecology of fish migration. *Global Ecology and Biogeography, 30*(1), 99–116.

Andrews, N., Bennett, N. J., Le Billon, P., Green, S. J., Cisneros-Montemayor, A. M., Amongin, S., & Sumaila, U. R. (2021). Oil, fisheries and coastal communities: A review of impacts on the environment, livelihoods, space and governance. *Energy Research & Social Science, 75*, Article 102009.

Andriesse, E., Saguin, K., Ablo, A. D., Kittitornkool, J., Kongkaew, C., Mang'ena, J., & Yang, J. (2022). Aligning bottom-up initiatives and top-down policies? A comparative analysis of overfishing and coastal governance in Ghana, Tanzania, the Philippines, and Thailand. *Journal of Rural Studies, 92*, 404–414.

Arafeh-Dalmau, N., Munguia-Vega, A., Micheli, F., Vilalta-Navas, A., Villaseñor-Derbez, J. C., Précoma-de la Mora, M., & Possingham, H. P. (2023). Integrating climate adaptation and transboundary management: Guidelines for designing climate-smart marine protected areas. *One Earth, 6*(11), 1523–1541.

Arthur, R. I., Skerritt, D. J., Schuhbauer, A., Ebrahim, N., Friend, R. M., & Sumaila, U. R. (2022). Small-scale fisheries and local food systems: Transformations, threats and opportunities. *Fish and Fisheries, 23*(1), 109–124.

AU Climate Change Strategy. (2022). African Union climate change and resilient development strategy and action plan. https://www.webberwentzel.com/News/Documents/2022/100722-AU-climate-change-strategy-2022.pdf

Baag, S., & Mandal, S. (2022). Combined effects of ocean warming and acidification on marine fish and shellfish: A molecule to ecosystem perspective. *Science of the Total Environment, 802*, Article 149807.

Bagsit, F. U., Frimpong, E., Asch, R. G., & Monteclaro, H. M. (2021). Effect of a seasonal fishery closure on sardine and mackerel catch in the Visayan Sea, Philippines. *Frontiers in Marine Science, 8*, Article 640772.

Bennett, M., March, A., & Failler, P. (2024). Blue economy financing solutions for the fisheries and aquaculture sectors of Caribbean Island states. *Fishes, 9*(8), 305.

Bernzen, A., Mangnus, E., & Sohns, F. (2022). Diversify, produce or buy? An analysis of factors contributing to household dietary diversity among shrimp and non-shrimp farmers in coastal Bangladesh. *Food Security, 14*(3), 741–761.

Bo, J., Zheng, R., Jiang, Y., Chen, J., Fang, C., Bailey, C., & Segner, H. (2023). Physiological-dependent alterations on transcriptomic and proteomic patterns of the single and combined temperature and salinity-exposed hybrid grouper, Epinephelus fuscoguttatus♀× Epinephelus lanceolatus♂. *Aquaculture, 575*, Article 739746.

Boncan, Y. M. L. (2021). *Casting hope for our collective future: Unpacking marine protection efforts of the Sulubaaï environmental foundation in Pangatalan Island, Taytay*. Ateneo De Manila University.

Boyd, C. E., McNevin, A. A., & Davis, R. P. (2022). The contribution of fisheries and aquaculture to the global protein supply. *Food Security, 14*(3), 805–827.

Brillo, B. B. C., & Simondac-Peria, A. C. (2021). Sustainability of a local government-instituted ecotourism development: Tayak adventure, nature and wildlife Park in Rizal, Laguna, Philippines. *Environment, Development and Sustainability*, 1–18.

Briscoe, D. K., Crowder, L. B., Balazs, G. H., Seminoff, J. A., Abreu, A., Lee Hing, C. A., & Polovina, J. J. (2025). Multi-decade northward shift of loggerhead sea turtle pelagic habitat as the eastern North Pacific Transition Zone becomes more oligotrophic. *Frontiers in Marine Science, 11*, 1513162.

Bryndum-Buchholz, A., Tittensor, D. P., & Lotze, H. K. (2021). The status of climate change adaptation in fisheries management: Policy, legislation and implementation. *Fish and Fisheries, 22*(6), 1248–1273.

Caballero, M. D., Gunda, T., & McDonald, Y. J. (2023). Energy justice & coastal communities: The case for meaningful marine renewable energy development. *Renewable and Sustainable Energy Reviews, 184*, Article 113491.

Chengula, F. I. (2023). *Knowing weather, strengthening livelihoods the role of indigenous knowledge weather forecasting in fishing communities in coastal Tanzania* (Doctoral dissertation, Doctoral Dissertation, University of Roehampton, London).

Cinti, A., Ramirez, L., Castrejón, M., Aburto, J. A., Loto, L., Fulton, S., & Parma, A. M. (2025). Small-scale fisheries in ecologically sensitive areas in Latin America and the Caribbean: Do marine protected areas benefit fisheries governance? *Ambio, 54*(1), 20–42.

Coates, D., McInnes, R. J., & Davidson, N. C. (2023). Linkages between inland fisheries and international instruments–opportunities for engagement: Food and Agriculture Organization of the United Nations Rome, 2023. *FAO Fisheries and Aquaculture Circular* (C1239), I-68.

Coger, T., Dinshaw, A., Tye, S., Kratzer, B., Aung, M. T., Cunningham, E., & Carthy, A. (2022). Locally led adaptation: From principles to practice. *World Resources Institute*, 1–32.

de Carvalho, D. A., Amaral, S., & Alves, L. M. (2023). Climate change adaptation frameworks in fishing communities: A systematic review. *Ocean & Coastal Management, 243*, Article 106754.

Dias, A. C. E., Armitage, D., Nayak, P. K., Akintola, S. L., Arizi, E. K., Chuenpagdee, R., & Singh, S. (2023). From vulnerability to viability: A situational analysis of small-scale fisheries in Asia and Africa. *Marine Policy, 155*, Article 105731.

Dimarchopoulou, D., Makino, M., Prayoga, M. R., Zeller, D., Vianna, G. M., & Humphries, A. T. (2022). Responses in fisheries catch data to a warming ocean along a latitudinal gradient in the western Pacific Ocean. *Environmental Biology of Fishes*, 1–16.

Ebarvia, M. C. M. (2022). *How well has environmental and social protection been ensured for small farmers and fisherfolk? Sustainable development of Philippine agriculture and fisheries* (No. 2022-11). PIDS Discussion Paper Series.

El-Sayed, A. F. M., & Fitzsimmons, K. (2023). From Africa to the world—The journey of Nile tilapia. *Reviews in Aquaculture, 15*, 6–21.

Evans, L. S., Buchan, P. M., Fortnam, M., Honig, M., & Heaps, L. (2023). Putting coastal communities at the center of a sustainable blue economy: A review of risks, opportunities, and strategies. *Frontiers in Political Science, 4*, 1032204.

Fischer, C., Trono, R. B., & Salinas, R. A. F. (2021). Addressing gaps in socio-economic and natural assets to halt marine turtle egg poaching: A livelihood feasibility study in the Turtle Islands Wildlife Sanctuary, Philippines. *Global Ecology and Conservation, 32*, Article e01913.

Gisevius, K., Niesters, L. M., & Braun, B. (2025). The role of community leadership in building community adaptive capacity to coastal hazards–Insights from neighborhood networks in Semarang, Indonesia. *Environmental Science & Policy, 163*, Article 103963.

Hilmi, N., Basu, R., Crisóstomo, M., Lebleu, L., Claudet, J., & Seveso, D. (2023). The pressures and opportunities for coral reef preservation and restoration in the Maldives. *Frontiers in Environmental Economics, 2*, 1110214.

Ho, T. Q., Do, H. L., & Eggert, H. (2025). Shrimp farming industry in Vietnam: An aquaculture performance indicators approach. *Aquaculture Economics & Management*, 1–24.

Islam, M. S., & Hasan, A. S. M. (2024). International legal arrangements for illegal, unreported, and unregulated fishing: Legal mechanisms and challenges in Bangladesh. *Marine Development, 2*(1), 1–18.

Jack-Kadioglu, T. (2022). *Power and polycentric coastal governance: Livelihoods, equity, and environmental change in the Philippines*. University of Exeter.

Jauharee, A. R. (2022). *The tuna pole and line FAD (fish aggregating device) fishery of the Maldives: Towards science-based management through fishers and scientific knowledge* (Doctoral dissertation, Université de Montpellier).

Jha, M. K., & Dev, M. (2024). Impacts of climate change. In *Smart internet of things for environment and healthcare* (pp. 139–159). Springer Nature Switzerland.

Kamau, J. N., Jacobs, Z. L., Jebri, F., Kelly, S., Kimani, E., Makori, A., & Painter, S. C. (2021). Managing emerging fisheries of the North Kenya Banks in the context of environmental change. *Ocean & Coastal Management, 209*, Article 105671.

Kim, H., Franco, A. C., & Sumaila, U. R. (2023). A selected review of impacts of ocean deoxygenation on fish and fisheries. *Fishes, 8*(6), 316.

Kirkman, S. P., Kowalski, P., Mann, B. Q., Branch, G. M., van der Bank, M. G., Sink, K. J., & Petersen, S. L. (2023). The road towards effective governance and management of marine protected areas in South Africa: Evolving policies, paradigms and processes. *African Journal of Marine Science, 45*(2), 63–86.

Kitolelei, S., Thaman, R., Veitayaki, J., Breckwoldt, A., & Piovano, S. (2021). Na Vuku makawa ni qoli: Indigenous fishing knowledge (IFK) in Fiji and the pacific. *Frontiers in Marine Science, 8*, Article 684303.

Kotowicz, D. M., Torell, E., Castro, J., Oracion, E. G., Pollnac, R., & Ricci, G. (2022). Exploring influences on environmental stewardship of fishing communities in fisheries management in the Philippines. *Environmental Management, 69*(6), 1102–1117.

Leotaud, N., Laidlow-Ferdinand, A., Clauzel, S., & Vaughan, N. (2024). Towards sustainable blue tourism in the Caribbean: Policy pathways to support community-based coastal and marine tourism. Blue Tourism Initiative Diagnostic Study. CANARI: Port of Spain. http://www.canari.org; www.bluetourisminitiative.org

Łukaszuk, T. (2024). Comparative study of maritime governance in the Global South—In Search for the Broader Cooperation. *Polish Political Science Yearbook, 53*(4), 141–164.

Mahmood, H., Ahmed, M., Islam, T., Uddin, M. Z., Ahmed, Z. U., & Saha, C. (2021). Paradigm shift in the management of the Sundarbans mangrove forest of Bangladesh: Issues and challenges. *Trees, Forests and People, 5*, Article 100094.

Manlosa, A. O., Hornidge, A. K., & Schlüter, A. (2021). Institutions and institutional changes: Aquatic food production in Central Luzon, Philippines. *Regional Environmental Change, 21*(4), 127.

March, A., & Failler, P. (2022). Small-scale fisheries development in Africa: Lessons learned and best practices for enhancing food security and livelihoods. *Marine Policy, 136*, Article 104925.

Matovu, B. (2024). Relevance of the high seas Treaty towards Ocean sustainability targets in the Global South. *KMI International Journal of Maritime Affairs and Fisheries, 16*(1), 21–54.

Miao, Q., & Nduneseokwu, C. (2025). Global environmental leadership: Addressing the triple planetary crisis. In *Environmental leadership in a VUCA era: An interdisciplinary handbook* (pp. 37–119). Springer Nature Singapore.

Mielgo, R., Gutiérrez, M., & Almazor, G. G. (2024). Fisheries depletion: The factors and actors involved in the transfer of fish from developing countries to developed ones. *Development*, 1–10.

Morales-Bojórquez, E., Nevárez-Martínez, M. O., García-Alberto, G., Villalobos, H., Aguirre-Villaseñor, H., Larios-Castro, E., & Martínez-Zavala, M. D. L. Á. (2021). Interaction between marine fauna and the small pelagic fishery in the coastal environment of the Gulf of California, México. *Frontiers in Marine Science, 8*, Article 669176.

Muringai, R. T., Mafongoya, P. L., & Lottering, R. (2021). Climate change and variability impacts on sub-Saharan African fisheries: A review. *Reviews in Fisheries Science & Aquaculture, 29*(4), 706–720.

Nagelkerken, I., Allan, B. J., Booth, D. J., Donelson, J. M., Edgar, G. J., Ravasi, T., & Mellin, C. (2023). The effects of climate change on the ecology of fishes. *PLOS Climate, 2*(8), Article e0000258.

Nakamura, J. N. (2022). Legal reflections on the small-scale fisheries guidelines: Building a global safety net for small-scale fisheries. *The International Journal of Marine and Coastal Law, 37*(1), 31–72.

Nakamura, J. N., Amador, T., Ibrahim, M., Nizar, H. R., & Haleem, M. (2023). *Legal report on the ecosystem approach to fisheries in Maldives: An analysis of the ecosystem approach to fisheries in selected national policy and legal instruments of Maldives*. Food & Agriculture Org.

Neokye, E., & Wang, X. (2024). Climate change and its impacts on marine fishery resources: A review. Available at SSRN 5008038. https://papers.ssrn.com/sol3/papers.cfm?abstract_id=5008038

Nyawade, O. B., Were-Kogogo, P., Adero, D. O., Owiti, P., Osimbo, H., & Chege, M. (2021). Dwindling fish catch in Kwale, Kenya: Vulnerable southern coast marine fisher communities and alternative sustainable livelihoods. *The International Journal of Community and Social Development, 3*(3), 255–273.

Obiero, K. O., Klemet-N'Guessan, S., Migeni, A. Z., & Achieng, A. O. (2023). Bridging Indigenous and non-Indigenous knowledge systems and practices for sustainable management of aquatic resources from East to West Africa. *Journal of Great Lakes Research, 49*, S128–S137.

Okeke-Ogbuafor, N., Taylor, A., Dougill, A., Stead, S., & Gray, T. (2022). Alleviating impacts of climate change on fishing communities using weather information to improve fishers' resilience. *Frontiers in Environmental Science, 10*, Article 951245.

Okyere, I., Chuku, E. O., Dzantor, S. A., Ahenkorah, V., & Adade, R. (2023). Capacity deficit and marginalisation of artisanal fishers hamper effective fisheries governance in Ghana: Insights and propositions for promoting sustainable small-scale fisheries. *Marine Policy, 153*, Article 105640.

Oloko, A., Harper, S., Fakoya, K., & Sumaila, U. R. (2024). The multi-dimensional perspectives of taboos on gender roles of fisherfolk in the Global South. *Maritime Studies, 23*(1), 1.

Omukuti, J., Barrett, S., White, P. C., Marchant, R., & Averchenkova, A. (2022). The green climate fund and its shortcomings in local delivery of adaptation finance. *Climate Policy, 22*(9–10), 1225–1240.

Onyango, H. O., Ochiewo, J. O., & Karani, N. J. (2021). Socio-economic prospects and problems in under-exploited offshore marine fisheries: The case of Fish Aggregating Devices (FADs) in Kenya coastal fisheries. *Regional Studies in Marine Science, 44*, Article 101706.

Paniagua, P., & Rayamajhee, V. (2024). Governing the global fisheries commons. *Marine Policy, 165*, Article 106182.

Philippines Fisheries Code. (1998). An act providing for the development, management and conservation of the fisheries and aquatic resources. https://www.bfar.da.gov.ph/wp-content/uploads/2021/02/Philippine-Fisheries-Code-of-1998.pdf

Riera, L. (2022). *From tensions to integrations: Development and conservation coalitions in Fijian coastal fisheries management* (Doctoral dissertation, Université Paul Valéry-Montpellier III; University of Bremen).

Roscher, M. B., Allison, E. H., Mills, D. J., Eriksson, H., Hellebrandt, D., & Andrew, N. L. (2022). Sustainable development outcomes of livelihood diversification in small-scale fisheries. *Fish and Fisheries, 23*(4), 910–925.

Sainjoo, S. (2024). *Adapting to uncertainties: Assessing the impacts of dams, Climate Change, and resource shifts on farming livelihoods in the Lower Mekong region* (Doctoral dissertation, Michigan State University).

Samanth, M. (2024). Ocean acidification and its impact on marine ecosystems. *Available at SSRN 4878195.*

Shanta, M. A., Suman, S. U., & Islam, M. M. (2025). Role of formal aquaculture industry for empowering women: A study in the southern coastal region of Bangladesh. In *Women's studies international forum* (Vol. 109, p. 103046). Pergamon.

Simon, J. W., & Muñoz, W. (2025). Making mainstreaming work for climate change adaptation: A multi-level analysis of adaptive policy capacity building in fishing and aquaculture governance in Chile. *Public Administration and Development*. https://doi.org/10.1002/pad.2098

Singhar, A. S., & Gundimeda, H. (2024). Sustainable Harvests under different bio-economic scenarios of Chilika wetland's fisheries. *Ecology, Economy and Society—the INSEE Journal, 7*(1), 75–103.

Sinthumule, N. I. (2023). Traditional ecological knowledge and its role in biodiversity conservation: A systematic review. *Frontiers in Environmental Science, 11*, 1164900.

Sobha, T. R., Vibija, C. P., & Fahima, P. (2023). Coral reef: A hot spot of marine biodiversity. In *Conservation and sustainable utilization of bioresources* (pp. 171–194). Springer Nature Singapore.

Sorensen, J. (1990). An assessment of Costa Rica's coastal management program. *Coastal Management, 18*(1), 37–63.

Umamaheswari, T., Sugumar, G., Krishnan, P., Ananthan, P. S., Anand, A., Jeevamani, J. J. J., & Rao, C. S. (2021). Vulnerability assessment of coastal fishing communities for building resilience and adaptation: Evidences from Tamil Nadu, India. *Environmental Science & Policy, 123*, 114–130.

Widjaja, S., Long, T., Wirajuda, H., As, H. V., Bergh, P. E., Brett, A., & Wilcox, C. (2023). Illegal, unreported and unregulated fishing and associated drivers. In *The Blue compendium: From knowledge to action for a sustainable ocean economy* (pp. 553–591). Springer International Publishing.

Zhang, K., Ye, Z., Qi, M., Cai, W., Saraiva, J. L., Wen, Y., & Zhao, J. (2025). Water quality impact on fish behavior: A review from an aquaculture perspective. *Reviews in Aquaculture, 17*(1), Article e12985.

CHAPTER 4

Pastoralist Livelihoods in the Face of Climate Change in the Global South: Climate Resilience Frameworks and Prospects

Abstract Pastoralists are among the most adversely impacted by climate change. This chapter explores the challenges and prospects of pastoralist livelihoods in the Global South in the context of climate change, focusing on resilience frameworks. Pastoralists in this region face significant challenges due to climate-induced droughts, land degradation, and the loss of grazing lands, which threaten their traditional way of life and food security. Despite these challenges, traditional resilience practices, such as transhumance, communal grazing land management, and the preservation of indigenous knowledge, continue to play a crucial role in sustaining pastoral systems. These practices allow communities to adapt to environmental variability, fostering resilience in the face of climate stress. Additionally, technological innovations such as mobile weather applications, early warning systems, and market access tools are increasingly supporting pastoralists by providing timely information for better decision-making, enhancing their capacity to mitigate climate risks. However, the sustainability of pastoralist systems remains uncertain, as climate change exacerbates existing pressures and requires continuous adaptation strategies. The future of pastoralist livelihoods in the Global South depends on their ability to diversify income sources, access new markets, and receive policy support to safeguard their practices. These prospects are vital for the long-term survival of pastoral communities,

N. P. Awazi, *Building Resilience*, Palgrave Studies in Climate Resilient Societies, https://doi.org/10.1007/978-3-031-93791-0_4

as they navigate the complexities of climate change while striving for sustainable and resilient livelihoods.

Keywords Pastoralists · Climate change · Resilience · Transhumance · Land degradation · Technology · Sustainability · Global South

4.1 Challenges Facing Pastoralists in the Global South

Pastoralism has been a central livelihood for many communities in the Global South, particularly in regions such as Sub-Saharan Africa, Central Asia, and parts of Latin America and the Middle East (Basupi, 2022; Manzano et al., 2021). Pastoralists are people who rely on livestock herding and grazing as their primary mode of subsistence. The relationship between these communities and their environment is symbiotic; however, various factors, particularly climate change, are increasingly challenging their way of life (McGuirk & Nunn, 2025). Among the most pressing challenges faced by pastoralists in the Global South are the impacts of droughts, land degradation, and the loss of grazing lands (Table 4.1). These challenges have been exacerbated by climate change, leading to food insecurity, loss of livelihoods, and displacement of communities. One of the most significant impacts of climate change on pastoralist communities is the increasing frequency and severity of droughts (Habte et al., 2022; Ndiritu, 2021). Pastoralists in arid and semi-arid regions depend heavily on the availability of water and grazing resources to sustain their livestock. Water scarcity is a critical issue, as animals require substantial amounts of water to survive, and without reliable sources, the pastoralists face dire consequences. Droughts, which are becoming more unpredictable due to climate change, can lead to severe livestock mortality, loss of income, and malnutrition. According to the Intergovernmental Panel on Climate Change (IPCC), the frequency of droughts has increased in many parts of the Global South, and this trend is expected to continue (IPCC, 2023). For example, in the Sahel region of Africa, a region known for its pastoralist communities, droughts are more frequent and prolonged. Devereux (2006) highlights that such climatic variations disrupt not only the immediate well-being of pastoralists but also the broader economic stability of their communities, as livestock is a central source of income, food, and cultural identity (Njoki, 2023; Tesfaye, 2022). In East Africa, Kenya's pastoralists, particularly in the northern and eastern regions, have faced significant challenges due to recurring droughts (Kirui et al., 2022; Yator, 2024). The 2016–2017 drought, which affected much of East Africa, resulted in widespread livestock deaths and the displacement of thousands of families. As climate change continues to exacerbate these conditions, it is increasingly difficult for pastoralists to predict when droughts will occur and how severe they will be (Tofu, 2024). This uncertainty reduces their ability to plan effectively, leading to diminished resilience.

Land degradation is another critical challenge for pastoralists, often linked to both human and environmental factors (Ntumva, 2022a, 2022b). Overgrazing, deforestation, and unsustainable farming practices exacerbate land degradation. In many parts of the Global South, pastoralists are confronted with diminishing soil fertility, reduced water retention capacity, and the encroachment of desertification (Azerigyik et al., 2024; Herrera Calvo, 2024). According to the United Nations Convention to Combat Desertification (UNCCD), desertification affects over 2 billion people worldwide, with pastoralist communities being particularly vulnerable. Land degradation reduces the availability of grazing land, making it harder for pastoralists to sustain their herds. Furthermore, soil erosion and the depletion of nutrients in the soil contribute to the long-term decline of productive land, leaving pastoralists with fewer resources to maintain their livelihoods. In Sub-Saharan Africa, the situation is dire. As noted by Aliyu and Sama'ila (2023), pastoralists in the Sahel region have witnessed a dramatic decline in the availability of grazing land due to land degradation and the encroachment of desertification. This phenomenon has been exacerbated by human-induced climate change, with unpredictable rainfall patterns and a reduction in overall precipitation. As the land becomes less fertile, pastoralists are forced to move further in search of adequate grazing resources, which leads to increasing competition over land and resources. In some cases, land degradation is coupled with population pressure and the expansion of agricultural land, which reduces the space available for traditional pastoralist practices. This phenomenon is particularly evident in regions where pastoralist communities are bordered by sedentary farming communities. As more land is converted to agricultural use, the traditional migratory routes of pastoralists become restricted, leading to conflicts over land and water resources.

Table 4.1 Challenges faced by pastoralists in the Global South

Challenge	*Description*	*Key examples*	*Key references*
Droughts	Increased frequency and severity of droughts due to climate change, leading to water scarcity, livestock mortality, and loss of income	• **Sahel region**: Increasing drought frequency and duration impacting pastoralist communities • **East Africa (Kenya)**: 2016–2017 drought causing widespread livestock deaths and displacement	Devereux (2006), Ndiritu (2021), Habte et al. (2022), McGuirk and Nunn (2025), Tesfaye (2022), Kirui et al. (2022)
Land degradation	Overgrazing, deforestation, and unsustainable farming practices lead to land degradation, reducing grazing land and soil fertility, exacerbating desertification	• **Sahel region**: Decline in grazing land due to land degradation and desertification • **Sub-Saharan Africa**: Diminishing soil fertility and reduced water retention capacity	Ntumva (2022a, 2022b), Azerigyik et al. (2024), UNCCD; Aliyu and Sama'ila (2023), Herrera Calvo (2024)
Loss of grazing lands	Shrinking grazing areas due to climate change, land fragmentation, and agricultural expansion, affecting mobility patterns and the socio-economic structure of pastoralist communities	• **East Africa**: Shrinking grazing lands leading to increased competition with farming communities • **Sub-Saharan Africa**: Traditional migration routes disrupted by agricultural expansion	Stavi et al. (2022), Boas (2022), Scoones (2021), Sani Ibrahim et al. (2021), Singh et al. (2022)
Water Scarcity	Reduced access to reliable water sources, critical for livestock survival, exacerbated by changing rainfall patterns and droughts	• **Arid and semi-arid regions**: Water scarcity severely impacting livestock and pastoralist livelihoods	Ndiritu (2021), IPCC (2023), Tofu (2024)

(continued)

Table 4.1 (continued)

Challenge	*Description*	*Key examples*	*Key references*
Marginalization and political factors	Political neglect of pastoralism, lack of recognition by governments, and conflicts over land and resources, exacerbated by policies prioritizing agriculture and urban development	• **Kenya's Rift Valley**: Conflict over grazing land and water resources. **Global South**: Policies favoring large-scale agriculture over pastoralist needs	Campbell (2022), Naess et al. (2025), Dutta Gupta et al. (2023), Dutta Gupta et al. (2023),
Cultural and social impacts	The reduction in grazing lands impacts social cohesion, traditional practices, and community stability	• **Global South**: Decline in traditional pastoralist practices, disrupting inter-community relationships and cultural connections	Scoones (2021), Little (2021)
Conflict over resources	Competition for dwindling resources such as land and water, exacerbated by climate change and population pressure, leading to violent conflicts	• **East Africa**: Displacement and violence over grazing land and water resources	Dutta Gupta et al. (2023), Tesfaye (2022), Kirui et al. (2022)
Livelihood vulnerability	As grazing lands shrink and livestock face higher mortality, pastoralists are increasingly forced to sell livestock, undermining long-term economic security and cultural identity	• **Sub-Saharan Africa**: Livestock sales during droughts lead to loss of family wealth and cultural symbols	Singh et al. (2022), Little (2021)
Unpredictable climate variability	Climate change introduces uncertainty regarding drought patterns, rainfall, and grazing resources, reducing the ability to plan and adapt effectively	• **East Africa**: Difficulty predicting the timing and severity of droughts, diminishing resilience	Kirui et al. (2022), Tesfaye (2022), Tofu (2024)

The loss of grazing lands is another critical challenge faced by pastoralists (Stavi et al., 2022). Climate change, land degradation, and the expansion of agriculture have all contributed to the shrinking of available grazing lands. As droughts become more frequent, pastoralists are forced to travel longer distances in search of water and pasture, disrupting their traditional grazing systems. The shifting availability of grazing lands due to climate change is altering the mobility patterns of pastoralists (Boas, 2022). Traditionally, many pastoralist communities have relied on transhumance or migratory practices, moving between different grazing areas depending on the season. However, the changing climate, land fragmentation, and the encroachment of farming communities make it increasingly difficult for pastoralists to maintain these traditional migration patterns (Sani Ibrahim et al., 2021). This has significant socio-economic and cultural consequences. The loss of grazing land not only affects the ability of pastoralists to sustain their livestock but also impacts the social fabric of pastoralist communities. In many cases, traditional grazing areas are sources of social and cultural connection, where inter-community relationships are forged. As these spaces diminish, there are often negative implications for community cohesion and social stability. Furthermore, the reduction of grazing lands forces pastoralists to make difficult decisions regarding their livestock (Scoones, 2021; Singh et al., 2022). With fewer grazing areas, the pressure to sell livestock increases, which undermines the long-term sustainability of their herding practices. Selling animals for immediate survival can lead to the loss of family wealth and a further weakening of pastoralist livelihoods, as livestock are not only an economic asset but also a cultural symbol (Little, 2021).

The challenges faced by pastoralists due to climate change are not only environmental but also deeply political (Campbell, 2022). The pastoralist way of life is increasingly marginalized in the Global South due to the political economy of land use, governance structures, and land tenure systems. Governments and international institutions often fail to recognize the value of pastoralism, which is viewed by many as an inefficient form of land use compared to settled agricultural practices. As highlighted by Naess et al. (2025), the marginalization of pastoralists in the political arena has led to policies that exacerbate their vulnerability to climate change. In many instances, governments prioritize large-scale agricultural development, mining, or urbanization over the needs of pastoralist communities. This policy neglect has led to the displacement of pastoralist communities from their land, leaving them more vulnerable to the impacts of climate change and environmental degradation. In some cases, pastoralists face violent conflict over access to land and resources. In Kenya, for instance, pastoralists in the Rift Valley have been involved in conflicts over grazing land and water resources, which are exacerbated by climate-induced scarcity (Dutta Gupta et al., 2023). These conflicts not only lead to loss of life and property but also further displace pastoralists, making it even harder for them to maintain their traditional way of life.

Hence, the challenges faced by pastoralists in the Global South are complex and multifaceted, driven by both environmental and political factors. Climate change has intensified droughts, water scarcity, land degradation, and the loss of grazing lands, putting immense pressure on pastoralist livelihoods. As these communities struggle to adapt to these changes, they face growing social, economic, and political marginalization. The lack of recognition of pastoralism's value by governments and international organizations further exacerbates the vulnerability of these communities. Addressing the challenges facing pastoralists requires comprehensive and inclusive policies that recognize the unique needs of pastoralist communities. This includes the protection of grazing lands, the promotion of sustainable land management practices, and the development of climate-resilient strategies that allow pastoralists to adapt to the changing climate. Furthermore, governments must engage with pastoralist communities to ensure that their voices are heard in the formulation of policies that affect their livelihoods. Without such measures, the survival of pastoralism as a livelihood and way of life in the Global South remains in jeopardy.

4.2 Traditional Resilience Practices of Pastoralists in the Global South

The pastoralists of the Global South, particularly in Sub-Saharan Africa, South Asia, and parts of Latin America, have long demonstrated impressive resilience in the face of environmental, economic, and social challenges (Fre et al., 2022). These resilient practices are deeply rooted in the cultural, ecological, and social fabric of these communities and have allowed them to adapt to a dynamic and often harsh natural environment (Moreda et al., 2024; Sharifian et al., 2022). Pastoralism, the practice of raising livestock in extensive, mobile systems, is not just an economic activity but also a way of life for millions of people. This resilience is underpinned by various traditional practices, including transhumance, communal grazing land management, and the integration of traditional ecological knowledge (Table 4.2). These practices not only sustain livelihoods but also foster social cohesion, biodiversity, and ecosystem health (Slayi et al., 2024).

Transhumance, the seasonal migration of livestock between fixed summer and winter pastures, is one of the most prominent resilience strategies of pastoralists in the Global South (Mbih et al., 2022; Neupane et al., 2022). This practice is particularly widespread in regions like the Sahel, the Horn of Africa, and the Andes. Transhumance allows pastoralists to spread the pressure of grazing across vast landscapes, reducing the risk of overgrazing and land degradation in any single area (Azimi et al., 2022; Behnke, 2021). By moving with the seasons, pastoralists can access different ecological zones that provide a diversity of resources at various times of the year. In the Sahel, for example, transhumance has been practiced for centuries by the Fulani and Tuareg pastoralists. They move their herds between the semi-arid zones of the Sahel and the more temperate zones of the southern savannahs or the mountain ranges of northern Mali and Niger. This seasonal mobility enables pastoralists to adapt to the unpredictable nature of rainfall and forage availability, key factors that determine the success of livestock production in arid and semi-arid regions. In the context of climate change, which has led to more erratic rainfall patterns, transhumance helps mitigate the risks of drought and pasture depletion by allowing animals to access better grazing areas during dry spells. Studies by Ouédraogo et al. (2021) and Timpong-Jones et al. (2023) have highlighted the role of transhumance in enhancing pastoral resilience. They argue that mobility is essential not only for resource access but also for maintaining genetic diversity in livestock populations, thereby strengthening the adaptability of pastoralist herds to changing environmental conditions. Furthermore, the practice fosters adaptive flexibility, allowing pastoralists to alter migration routes and grazing patterns based on evolving environmental cues, such as the onset of rains or the availability of water.

Communal land management is another cornerstone of resilience among pastoralist communities (Abdela, 2024; Senda et al., 2022). Traditional grazing systems are often governed by collective rules and norms, which are designed to prevent overuse of common resources and ensure equitable access to grazing land and water. This system fosters both environmental sustainability and social cohesion, as grazing lands are seen

Table 4.2 Traditional resilience practices of pastoralists in the Global South

Resilience practice	*Description*	*Key examples*	*Key references*
Transhumance (seasonal migration)	Seasonal migration of livestock between fixed summer and winter pastures, allowing pastoralists to spread grazing pressure and access diverse resources based on environmental cues	• **Sahel (Fulani, Tuareg)**: Mobility between semi-arid and temperate zones, adapting to unpredictable rainfall and forage availability • **Horn of Africa**: Seasonal movement to access better grazing during dry spells	Neupane et al. (2022), Mbih et al. (2022), Behnke (2021), Azimi et al. (2022), Ouédraogo et al. (2021), Timpong-Jones et al. (2023)
Communal land management	Collective management of grazing lands, governed by local norms and decision-making processes to prevent overuse, ensure equity, and promote sustainability	• **Maasai (East Africa)**: Resource-sharing, rotating grazing areas, and cooperative herding to manage land and prevent overgrazing • **Quechua and Aymara (Andes)**: Ayllu system for collective land use and resource management	Senda et al. (2022), Abdela (2024), Flintan et al. (2021), Robinson and Flintan (2022), McKay (2023), Van Klingeren and de Graaf (2021)
Traditional Ecological Knowledge (TEK)	Deep knowledge of local ecosystems, weather patterns, animal husbandry, and land management passed down through generations, enabling adaptive responses to environmental changes	• **Somali (Horn of Africa)**: Use of subtle environmental indicators (e.g., animal behavior, plant species) to predict droughts and rainfall patterns • **Nepali (Himalayas)**: High-altitude grazing management to prevent overgrazing in fragile mountain ecosystems	Xie et al. (2024), Nulkar (2024)
Selective breeding for resilience	Breeding livestock that are better adapted to local environmental conditions, such as drought resistance or disease tolerance, ensuring herd survival during harsh environmental conditions	• **Nguni Cattle (Southern Africa)**: Known for drought resistance, selectively bred for resilience in harsh environments (Sharifian et al., 2022)	Sharifian et al. (2022)

(continued)

Table 4.2 (continued)

Resilience practice	*Description*	*Key examples*	*Key references*
Conflict resolution and resource-sharing	Systems for resolving conflicts over grazing rights and ensuring fair access to shared resources like land and water, fostering cooperation among pastoralist groups	• **Maasai (East Africa)**: Elders mediate grazing rights and manage resources to prevent conflict and overuse	Flintan et al. (2021), Robinson and Flintan (2022)
Integration of modern tools with traditional practices	The use of technology (e.g., mobile phones, satellite imagery) alongside traditional practices to monitor environmental changes, manage migrations, and improve resilience	• **Global South**: Integration of mobile phones and satellite imagery to predict weather patterns and manage livestock movement	Sharifian et al. (2022), Nori and Scoones (2023)
Sustainable resource use	Practices that balance resource extraction with regeneration, ensuring that land, water, and grazing areas remain productive over time	• **Quechua and Aymara (Andes)**: Rotation of grazing areas to allow pastures to regenerate and prevent soil degradation	Van Klingeren and de Graaf (2021), McKay (2023)

as a shared resource that requires communal responsibility and collective decision-making. In East Africa, for example, the Maasai pastoralists have developed elaborate systems of resource-sharing and conflict resolution related to grazing rights (Flintan et al., 2021; Robinson & Flintan, 2022). Traditionally, land and water access were regulated by elders and local leaders, who made decisions based on the availability of pasture and water, ensuring that resources were not depleted by overgrazing. This system also involved rotating grazing areas to give pastures time to regenerate, thus preventing soil erosion and maintaining biodiversity. The Maasai and other pastoral communities in the region have often engaged in cooperative herding, where different families and clans share the risks and rewards of livestock production, enhancing resilience in the face of environmental uncertainty. In the Andean highlands of South America, the indigenous Quechua and Aymara peoples have practiced similar forms of communal land management for centuries. The concept of *ayllu*, a traditional form of collective organization, has allowed pastoralists to share grazing lands and water sources, ensuring that these resources are not depleted by individual interests. The *ayllu* system also involves rotation of grazing areas and the careful management of natural resources to prevent degradation. These practices, which have evolved over generations, are integral to maintaining the productivity of the land and the health of livestock. Studies by McKay (2023), Van Klingeren and de Graaf (2021), and Agrawal et al. (2023) have argued that communal management systems are particularly effective in managing common-pool resources, where individual ownership could lead to overexploitation. These systems encourage collaboration and mutual support, allowing pastoralist communities to withstand external pressures such as land encroachment, climate change, and conflicts over land use.

One of the most profound aspects of pastoral resilience is the integration of traditional ecological knowledge (TEK), which includes deep understanding of local ecosystems, animal husbandry practices, weather patterns, and land management techniques (Xie et al., 2024). Pastoralists possess a rich repository of knowledge, passed down through generations, that enables them to interpret and respond to environmental changes in adaptive ways. This knowledge is not static but rather evolves in response to changing environmental conditions, making it a dynamic and flexible tool for resilience. For instance, in the Horn of Africa, Somali pastoralists are known for their ability to predict droughts and rainfall patterns through subtle indicators such as the behavior of certain animals, the appearance of certain plant species, or the wind direction. This knowledge allows pastoralists to plan their migrations and manage their herds effectively. Similarly, in the Himalayas, the high-altitude pastoralists of Nepal use their understanding of alpine ecosystems to manage grazing patterns and avoid overgrazing in fragile mountain pastures. TEK also encompasses the management of livestock breeds. Many pastoralist communities selectively breed animals that are more resilient to local environmental conditions, such as droughts, diseases, and extreme temperatures. These traditional breeding practices have led to the development of livestock breeds that are highly adapted to specific ecological niches. For example, the Nguni cattle of southern Africa are known for their drought resistance and ability to thrive in harsh environments, a result of centuries of selective breeding by pastoralists. The importance of TEK in maintaining resilience is well-documented in the work of Nulkar (2024) who emphasizes that traditional knowledge systems are often more effective in managing complex ecological processes than modern scientific approaches, particularly in resource-poor areas where access to technology and formal education may be limited.

While traditional resilience practices have been highly effective, they are increasingly under threat due to modernization, land privatization, and the impacts of climate change. The encroachment of agricultural lands, infrastructure development, and government policies that prioritize sedentary farming over mobility have led to the fragmentation of grazing lands and the erosion of communal management systems (Nori & Scoones, 2023). Climate change exacerbates these challenges, as changing rainfall patterns, rising temperatures, and increased frequency of extreme weather events undermine the predictability of traditional environmental cues. Despite these challenges, many pastoralist communities are finding ways to adapt and integrate modern strategies with traditional practices (Sharifian et al., 2022). In some cases, this includes the use of mobile phones and satellite imagery to monitor weather patterns and coordinate migrations. Furthermore, initiatives such as the International Fund for Agricultural Development's (IFAD) pastoralist development programs aim to support and strengthen traditional pastoralist systems by promoting sustainable land management, enhancing access to markets, and fostering policies that recognize the rights of pastoralists.

The resilience of pastoralists in the Global South is therefore built on a foundation of traditional practices that have evolved over centuries in response to specific ecological and social conditions. Transhumance, communal grazing land management, and traditional ecological knowledge are central to this resilience, allowing pastoralist communities to adapt to environmental variability, protect biodiversity, and maintain social cohesion. While these practices are increasingly under threat from modernization and climate change, they remain a powerful testament to the ingenuity and adaptability of pastoralist societies. Efforts to support and preserve these traditional practices, while integrating them with modern tools and technologies, offer a promising path toward sustainable pastoralism in the face of contemporary challenges.

4.3 Technological Innovations and Climate Resilience of Pastoralists in the Global South

The pastoralist communities in the Global South are facing increasingly erratic and extreme climate events, which are disrupting their livelihoods, food security, and socio-cultural practices (Fonchingong Che & Bang, 2024). Climate change exacerbates the challenges already faced by these communities, such as droughts, floods, and unpredictable rainfall patterns. However, the integration of modern technologies, such as mobile weather apps, early warning systems, and market access tools, is emerging as a transformative means for pastoralists to adapt to these challenges and build resilience (Table 4.3).

Pastoralists, who rely primarily on livestock for their livelihood, are particularly vulnerable to the impacts of climate change. These communities, many of which reside in arid and semi-arid regions of Africa, Asia, and Latin America, face challenges related to erratic rainfall, prolonged droughts, and increasing temperatures. Livestock, being highly sensitive to climatic shifts, suffers from dehydration, malnutrition, and disease outbreaks, which negatively impact productivity and food security (Lumborg et al., 2021). Additionally, pastoralist communities often face limited access to resources such as water, pasture, and markets, making their livelihoods even more precarious in the face of climate change (Tugjamba et al., 2023). Traditional coping mechanisms, such as mobility and diversification of grazing strategies, have been strained by the intensifying climate crisis. As a result, the need for adaptive strategies that leverage modern technology has become critical. Technological innovations offer the potential to provide pastoralists with real-time information, predictive analytics, and improved access to markets and services, thus enhancing their ability to respond to climate-related shocks (Banerjee et al., 2023).

One of the most significant technological innovations in recent years has been the development and adoption of mobile weather apps. These apps, which provide real-time weather forecasts, long-term climate predictions, and localized weather patterns, have proven essential for pastoralists in making informed decisions about grazing, water access, and animal management. A study by Turnbull and Harrison (2024) highlighted that mobile-based weather services help pastoralists plan for the arrival of droughts or excessive rainfall, allowing them to relocate livestock to more favorable areas and avoid potential loss. The information provided by

Table 4.3 Technological innovations and climate resilience of pastoralists in the Global South

Technological innovation	*Description*	*Key examples*	*Key references*
Mobile weather apps	Mobile applications that provide real-time weather forecasts, climate predictions, and localized weather patterns to assist pastoralists in decision-making regarding grazing and animal management	• **iCow (Kenya)**: Provides weather alerts and climate data to assist pastoralists in planning livestock movements • **Kenya meteorological department's app**: Offers timely weather updates and alerts for pastoralists	Turnbull and Harrison (2024), Daum et al. (2022), Fonchingong Che and Bang (2024)
Early warning systems (EWS)	Systems that combine meteorological data, satellite imagery, and local knowledge to issue warnings about climate events such as droughts, floods, and storms, enabling proactive responses	• **Horn of Africa**: Integrated early warning systems providing advance notifications of droughts or extreme weather events • **Kenya meteorological department's EWS**: Distributes alerts via mobile messages	Kaguai (2023), Otieno et al. (2024), Mabumbo et al. (2024), Ouko (2024)
Mobile-based market access	Technologies such as mobile phones, mobile money platforms, and e-commerce tools that improve pastoralists' ability to access markets and conduct transactions without needing physical infrastructure	• **M-Pesa (Kenya)**: Mobile money platform enabling pastoralists to send and receive payments, access loans, and engage in cash transfers • **Mobile market platforms**: Allow pastoralists to access real-time pricing and sell livestock directly to buyers	Parlasca (2021), Makokha et al. (2022), Duale (2024), Tuwei and Tully (2021)

(continued)

Table 4.3 (continued)

Technological innovation	*Description*	*Key examples*	*Key references*
Mobile Health (mHealth) technologies	Mobile apps and SMS-based services providing pastoralists with information on livestock health, disease prevention, treatments, and breeding strategies to improve livestock productivity and resilience	• **Waterpoints Monitoring platform (Ethiopia)**: Supports pastoralists in managing water resources and adapting to climatic challenges • Mobile-based veterinary services for disease management and prevention	Omondi et al. (2021), LD4D (2024)
Livestock health management platforms	Platforms enabling pastoralists to manage animal diseases, access treatment advice, and monitor herd health, reducing livestock mortality during extreme climate events	• **Ethiopia**: Platforms that offer veterinary advice, disease control strategies, and breeding recommendations for resilience • **SMS-based veterinary services**: Information on livestock health management	Omondi et al. (2021)
Digital financial services	The use of mobile money and financial tools that support pastoralists in managing finances, accessing loans, and engaging in trade, thus increasing economic resilience to climate shocks	• **M-Pesa (Kenya)**: Revolutionized financial transactions for pastoralists, enabling them to buy, sell, and manage funds digitally	Tuwei and Tully (2021)

(continued)

Table 4.3 (continued)

Technological innovation	*Description*	*Key examples*	*Key references*
Market information platforms	Platforms that offer real-time pricing information, helping pastoralists make informed decisions about when and where to sell livestock, improving economic resilience amid climate shocks	• **Mobile-based pricing platforms**: Help pastoralists avoid price volatility by providing timely market information	Parlasca (2021), Makokha et al. (2022)
Barriers to technology adoption	Challenges such as digital divides and limited access to infrastructure that hinder the widespread adoption of technologies among pastoralists	• **Remote areas**: Limited access to smartphones, internet, and digital services in isolated regions, especially due to political instability or infrastructure gaps	Xiandong (2023), Rigby et al. (2023)

these apps helps communities predict not only the weather but also potential health risks to their livestock, such as diseases that spread during periods of extreme weather conditions. In Kenya, mobile apps such as "iCow" and "Kenya Meteorological Department's weather app" have made significant strides in providing pastoralists with timely weather alerts, enabling them to take preemptive actions to safeguard their herds (Daum et al., 2022). Through these tools, pastoralists can access forecasts and climate data to make informed decisions about when to move their animals, thus improving livestock survival rates and reducing the economic impact of climate-induced shocks.

Early warning systems (EWS) have become a critical tool in enhancing the resilience of pastoralist communities to climate-related disasters. These systems rely on a combination of meteorological data, satellite imagery, and local knowledge to issue timely warnings about upcoming climate events, such as droughts, floods, and storms. In the Horn of Africa, early warning systems have been integrated with mobile technologies to provide pastoralists with advance notifications of impending droughts or extreme weather events (Kaguai, 2023; Otieno et al., 2024).

For instance, the Integrated Food Security Phase Classification (IPC) system provides real-time updates on food security risks, which are linked with climate patterns (IPC, 2024). Similarly, the "Kenya Meteorological Department" offers early warning information through text messages, which can be disseminated to pastoralists and other rural communities via mobile phones. These EWS allow pastoralists to take timely action, such as moving livestock to safer areas, stockpiling water, or engaging in emergency sales of animals to raise funds for food and other necessities. Early warning systems also help reduce the risks associated with delayed response to climate-induced disasters, thereby mitigating the negative consequences of droughts and floods on pastoralist communities (Mabumbo et al., 2024; Ouko, 2024).

In addition to providing weather information and early warnings, technological innovations also play a crucial role in improving pastoralists' access to markets (Duale, 2024; Makokha et al., 2022; Parlasca, 2021). Mobile phones, mobile money platforms, and e-commerce tools have significantly enhanced pastoralists' ability to sell their livestock and related products, even in remote areas. This market access is essential not only for increasing income but also for accessing critical services such as veterinary care, feed, and water. Mobile money services, such as M-Pesa in Kenya, have revolutionized the way pastoralists conduct business, allowing them to send and receive payments, access loans, and engage in cash transfers without needing physical bank infrastructure (Tuwei & Tully, 2021). Through these services, pastoralists can sell livestock or livestock products directly to buyers without having to travel long distances to urban markets, which are often located far from pastoralist areas. This helps reduce transportation costs, mitigates the risks of market failure, and ensures that pastoralists receive fair prices for their products, thereby enhancing their economic resilience. Moreover, mobile-based market platforms enable pastoralists to access real-time pricing information, allowing them to make informed decisions about when and where to sell their livestock. Through the use of these platforms, pastoralists can reduce the negative impact of price volatility, which is often exacerbated by climate-related factors such as droughts or disease outbreaks. These tools not only improve income generation but also enable pastoralists to engage in value-added activities, such as the processing and marketing of dairy or meat products, thereby diversifying their sources of income.

Another area where technology is making a significant impact is in the management of livestock health. Mobile health (mHealth) technologies, which include mobile apps, SMS-based services, and telemedicine platforms, are increasingly being used to provide pastoralists with information on animal diseases, treatments, and prevention strategies (Omondi et al., 2021). In Ethiopia, the "Waterpoints Monitoring platform" platform helps build the resilience of vulnerable pastoralist communities, supporting them in adapting to rising climatic challenges (LD4D, 2024). These services, which can be accessed through smartphones or feature phones, enable pastoralists to manage livestock diseases more effectively, reducing the risk of epidemics that can decimate herds during periods of climatic stress. By using mobile-based veterinary services, pastoralists can also improve livestock productivity, as healthier animals are more likely to thrive in challenging climates. In addition, these technologies offer vital support for animal breeding programs, allowing pastoralists to select for climate-resilient traits such as heat tolerance, drought resistance, and disease resistance, further bolstering their resilience to climate variability.

Despite the positive contributions of technology to pastoralist resilience, several challenges limit their widespread adoption. One of the primary barriers is the digital divide, which affects rural and remote communities. While mobile phone penetration has increased in many parts of the Global South, there are still significant gaps in access to smartphones, internet connectivity, and the necessary infrastructure to support digital services. This is particularly true in areas that are geographically isolated or affected by political instability (Rigby et al., 2023; Xiandong, 2023). Additionally, there is a lack of local capacity to fully utilize these technologies. Many pastoralists may lack the technical skills or literacy to interpret weather forecasts, use mobile apps effectively, or access market platforms. Therefore, it is essential to complement technological solutions with training programs and support systems that enhance the digital literacy of pastoralist communities.

4.4 Sustainability of Pastoral Systems in the Global South in the Face of Climate Change

Pastoralism, a livelihood based on the rearing of livestock on extensive rangelands, is a critical component of agricultural and economic systems in many regions of the Global South (Nori & Scoones, 2023). From the drylands of Africa to the steppes of Central Asia and the Andean highlands, pastoral systems are not just means of subsistence, but are also deeply interwoven with the culture, social organization, and political life of these regions. In the face of escalating climate change, the survival of pastoralist systems has come under increasing pressure, as more frequent and severe droughts, changes in precipitation patterns, and extreme weather events challenge traditional ways of managing livestock and natural resources (Aderinoye-Abdulwahab & Ganiyu, 2025; Akinyetun & Ogunbodede, 2023). Despite these challenges, pastoralist systems have demonstrated a remarkable capacity to adapt, both historically and in contemporary contexts, to a changing climate. Drawing on long-standing adaptive strategies, knowledge systems, and social practices, many pastoralist communities continue to thrive in the Global South.

The impacts of climate change on pastoral systems are multifaceted, with the most immediate threats arising from changing precipitation patterns, more frequent droughts, and temperature extremes. These changes lead to a direct reduction in the availability and quality of pasturelands, which are vital for the health of livestock. According to Thornton et al. (2009), the projected increases in temperature and shifts in rainfall patterns could significantly decrease the productivity of grazing lands, especially in semi-arid and arid regions, which are home to many pastoralist communities (Habte et al., 2022; Tugjamba et al., 2023). In the Sahel, for example, erratic rainfall and increasing temperatures have led to a reduction in the availability of grazing land, forcing pastoralists to travel longer distances in search of pasture and water. This has put immense pressure on the mobility of pastoral communities, which traditionally rely on their ability to move with their herds to access resources. Furthermore, the impacts of climate change are exacerbated by ongoing land degradation, deforestation, and population pressures, which limit the land available for grazing. The effect of climate change is also felt in the form of increased competition for resources between pastoralists and other land users, such as farmers, who are also vulnerable to climate-induced crop failures. These resource conflicts can lead to tensions and violence, as observed in parts of Sub-Saharan Africa, such as in the conflicts between pastoralists and farmers in the Horn of Africa.

Despite the considerable challenges posed by climate change, pastoral systems in the Global South have shown an impressive ability to adapt to shifting environmental conditions. Pastoralists have developed a variety of coping mechanisms over the centuries that have allowed them to navigate the uncertain and variable climates of their regions. These strategies include mobility, diversification, traditional knowledge systems, and the use of flexible governance structures.

The most fundamental adaptive strategy of pastoralists is mobility (Sulieman & Young, 2023). Pastoralists are often able to move their herds to areas where pasture and water are available, and this mobility allows them to avoid areas affected by drought or other adverse climate events. This flexibility in land use is one of the primary reasons pastoralism has been able to survive in harsh climates for millennia (Ndiritu, 2021). However, mobility is increasingly under threat due to land tenure systems that restrict movement, as well as the growing competition for land and water resources from sedentary farming communities and urbanization. Nevertheless, in many regions, pastoralist communities continue to use traditional migration routes that are based on centuries of knowledge of climate and weather patterns. In northern Kenya, for example, the Borana pastoralists have developed sophisticated systems of communal resource management and seasonal movement strategies that allow them to adapt to seasonal fluctuations in resource availability (Mutisya-Kioko, 2024).

Diversification is another key strategy employed by pastoralists to reduce their vulnerability to climate risks. While livestock remains the primary asset, many pastoralists also engage in crop cultivation, trade, and other forms of income generation. This diversification spreads risk across multiple sectors and reduces the likelihood of total failure in any one area (Ahmed et al., 2023; Sewando, 2022; Tofu et al., 2025). In some cases, pastoralists also engage in agro-pastoralism, combining pastoralism with crop farming in more fertile regions or in areas where rainfall is more reliable. This allows pastoral communities to buffer against the impacts of drought or poor grazing conditions.

Pastoralist communities have developed intricate systems of knowledge that are passed down through generations, enabling them to understand and predict environmental changes (Rajput, 2024). This traditional ecological knowledge (TEK) includes knowledge of weather patterns, the behavior of animals, and the characteristics of different plant species. In many cases, pastoralists are highly skilled at recognizing early signs of drought or other climatic changes and can adjust their practices accordingly. For instance, the Maasai people of East Africa have developed complex systems for managing grazing lands, based on the understanding that certain plants and grasses are more resilient to drought than others. Similarly, the Himba people in Namibia use a deep knowledge of weather patterns and animal behavior to time their movements and grazing activities (Inman, 2024). This knowledge system is vital for adapting to climate variability, and efforts to integrate it into modern climate adaptation strategies are critical for ensuring the sustainability of pastoralism.

Pastoralist communities have developed flexible governance structures that allow them to adapt to changing circumstances (Duale, 2024; Zhang, 2023). These systems are often based on communal decision-making and local-level resource management. In many cases, pastoralist groups rely on customary laws and practices to regulate access to grazing lands and water resources, ensuring equitable distribution during periods of scarcity. These flexible governance structures are important because they enable rapid responses to environmental changes. In times of crisis, pastoralists often rely on social networks for mutual support. Kinship and clan-based systems of sharing resources such as water, pasture, and livestock provide a safety net for vulnerable individuals within the community. This resilience has been observed in the Borana pastoralist community in Ethiopia, where strong social networks help mitigate the impacts of drought (Birhanu et al., 2021; Tofu et al., 2023).

While pastoral systems possess considerable adaptive capacity, several challenges threaten their long-term sustainability in the face of climate change with the most prominent being policy and land tenure issues, increasing conflict, and economic pressures (Ng'ang'a, 2022). One of the primary challenges facing pastoralists today is the loss of access to land and the erosion of mobility due to land tenure systems that favor sedentary farming. In many parts of the Global South, pastoralists are being pushed off their traditional grazing lands by state policies, commercial agriculture, and urbanization. This loss of land reduces the ability of pastoralists to move their herds to new grazing areas, making them more vulnerable to climate-related shocks. The lack of secure land tenure for pastoralists also makes it difficult for them to invest in sustainable land management practices. Many governments fail to recognize the value of mobile pastoralism, and policies often prioritize agricultural land over pastoral grazing lands. As a result, pastoralists are increasingly being marginalized, and their traditional systems of governance and resource management are undermined (Barrow & Fellow, 2022; Ntumva, 2022a, 2022b). Resource competition between pastoralists and farmers, as well as between different pastoralist groups, is escalating as climate change increases the scarcity of grazing lands and water. In regions like the Sahel, competition for land and water is leading to frequent conflicts between pastoralists and farmers, often with devastating consequences. Climate change exacerbates these conflicts, making it more difficult for pastoralists to maintain their livelihoods. The economic pressures facing pastoralists are also significant. In many regions, pastoralism is being increasingly commercialized, and many pastoralist communities are being integrated into national and global markets. While this can provide opportunities for income diversification, it also exposes pastoralists to the volatility of commodity prices and global market fluctuations, which are compounded by climate variability. As a result, pastoralist livelihoods are becoming more precarious.

Thus, the sustainability of pastoral systems in the Global South is threatened by climate change, but these systems also have significant potential for adaptation. The mobility, diversification, traditional knowledge systems, and flexible governance structures that have sustained pastoralism for centuries provide a foundation for resilience in the face of climate risks. However, these adaptive capacities are increasingly undermined by policies that restrict mobility, land tenure issues, and the erosion of social networks. To ensure the long-term sustainability of pastoral systems, it is crucial that governments, development agencies, and the international community recognize the value of pastoralism and support policies that enhance the resilience of pastoralist communities. This includes securing land tenure, integrating traditional knowledge into climate adaptation strategies, and fostering cross-sectoral collaboration to address resource conflicts. By empowering pastoralists and ensuring that their adaptive strategies are supported, it is possible for these systems to continue to thrive in the face of a changing climate.

4.5 Prospects for the Future in Pastoralist Communities in the Global South in the Face of Climate Change

Pastoralism, an ancient way of life based on the breeding and movement of livestock, has sustained millions of people in the Global South for centuries. Pastoralist communities, spread across Sub-Saharan Africa, the Middle East, and parts of Asia, have long relied on this mobile system to adapt to the harsh and often unpredictable environments they inhabit. However, the future of pastoralism faces a growing set of challenges, particularly in the context of climate change, which is intensifying environmental stresses, such as droughts, shifting rainfall patterns, and the degradation of grazing land. Despite these challenges, pastoralist communities are not without hope. There are opportunities to ensure their survival and thrive in the face of climate change. The future of these communities depends largely on how they adapt, diversify their livelihoods, access supportive policies, and engage with global markets. The prospects for the future of pastoralist communities in the Global South depends on the potential for sustainable livelihoods, diversification, policy support, and global market integration.

4.5.1 Climate Change and Its Impact on Pastoralism

Climate change is already affecting pastoralist communities in numerous ways. In regions like the Horn of Africa, Central Asia, and the Sahel, rising temperatures, prolonged droughts, and erratic rainfall patterns have severely impacted pasture availability, water sources, and the health of livestock. These climatic shifts lead to reduced grazing land, more frequent and intense water shortages, and increased competition for resources. As a result, pastoralists are finding it harder to sustain their herds and livelihoods. Moreover, climate-induced stresses often exacerbate existing socio-economic vulnerabilities, including poverty, food insecurity, and conflicts over land and resources. For example, droughts can lead to the loss of livestock, which is not only a financial blow but also a cultural loss, as livestock hold significant value beyond economic terms in many pastoralist communities. The cumulative effects of climate change are pushing pastoralist societies to rethink their survival strategies and consider ways to adapt and diversify their livelihoods.

4.5.2 *Diversification of Livelihoods*

One of the key strategies for ensuring the future of pastoralist communities in the face of climate change is livelihood diversification. Diversification allows pastoralists to reduce their dependency on traditional pastoralism and develop alternative sources of income. This strategy can include various activities such as agriculture, trade, tourism, and even renewable energy production. Pastoralists can complement livestock farming with crop production. In many regions, pastoralists have historically practiced mixed farming, combining livestock with crop production to ensure food security. In areas where water availability is becoming more unpredictable, cultivating drought-resistant crops such as millet, sorghum, or quinoa can help provide a buffer against the effects of climate change. Additionally, agroforestry practices such as planting trees alongside crops, can help mitigate soil erosion, enhance biodiversity, and improve water retention in the soil. Diversification within livestock production itself can also enhance resilience. For example, some pastoralists may choose to diversify the types of animals they raise, adding goats, poultry, or camels to their herds. These animals are often more adaptable to climate extremes, such as extreme heat or drought. Diversifying the genetic stock of livestock through breeding programs can also increase resilience to diseases and changing environmental conditions. In addition to agriculture, pastoralists can explore non-livestock-based livelihoods. This may include engaging in handicrafts, weaving, or other artisanal products that can be sold locally or internationally. For example, pastoralist communities in East Africa have successfully created businesses based on traditional leather goods, which not only generate income but also preserve cultural heritage. Similarly, community-based tourism, where visitors come to experience the life of pastoralists, can provide a significant income stream. These ventures also provide a platform for cultural exchange and can foster greater awareness of the challenges faced by pastoralists. Another promising diversification opportunity is the development of renewable energy sources, such as solar or wind energy. In many remote pastoralist communities, access to electricity is limited or nonexistent. The installation of solar panels for households or community use can reduce energy poverty and create new business opportunities. Pastoralists can also engage in the sale of excess energy, providing a supplemental income stream.

4.5.3 *Policy Support and Enabling Environments*

Effective policy support is crucial for the future of pastoralist communities. Governments in the Global South must recognize the importance of pastoralism as both a livelihood and a cultural practice. This includes providing policies that facilitate access to land, water, and markets, as well as fostering a supportive regulatory environment. One of the primary concerns for pastoralists is secure access to land and water. In many parts of the Global South, land tenure systems are poorly defined, and pastoralists often face encroachment on their grazing lands due to urbanization, agricultural expansion, and infrastructure development. As climate change intensifies, the pressure on land and water resources increases, leading to conflicts over access. Governments need to implement policies that secure land tenure for pastoralists, protecting them from displacement and enabling them to make long-term investments in their livelihoods. Pastoralist communities also need access to community-based resource management systems. By involving pastoralists in the governance of land and water resources, policymakers can ensure that local knowledge and practices are integrated into sustainable management efforts. This participatory approach can enhance the resilience of pastoral communities to climate change and ensure that their traditional knowledge of resource use is preserved. Governments can also support climate adaptation strategies that directly benefit pastoralists. For instance, governments can help pastoralists adopt climate-resilient agricultural practices, such as early warning systems for droughts, crop insurance programs, and support for livestock vaccination campaigns. International cooperation through initiatives such as the UN's Sustainable Development Goals (SDGs) can also play a key role in providing funding for climate adaptation projects tailored to the needs of pastoralist communities.

4.5.4 *Access to Global Markets*

In today's interconnected world, pastoralists in the Global South can benefit from access to global markets for their livestock and related products. However, this is often easier said than done. To fully capitalize on these opportunities, pastoralists need support to navigate the complexities of international trade, including complying with export standards and accessing international supply chains. Governments and international organizations should create infrastructure to facilitate trade between pastoralist communities and global markets. This includes improving roads, cold storage facilities, and market access for pastoral products, as well as establishing fair trade certifications that can help pastoralists fetch higher prices for their goods. Additionally, governments can promote the export of livestock products such as leather, wool, and milk, which are highly valued in global markets. The rise of digital platforms and mobile technologies offers new avenues for pastoralists to connect with global markets. Mobile phones and internet access allow pastoralists to access market information, weather forecasts, and prices for their products. Mobile banking and payment systems can also enhance financial inclusion, allowing pastoralists to engage in secure transactions and access credit for investment in their businesses.

Hence, the future of pastoralist communities in the Global South in the face of climate change will largely depend on their ability to adapt, diversify, and access supportive policies and markets. Climate change presents significant challenges, but it also offers opportunities for innovation and resilience. Livelihood diversification, including agricultural and non-agricultural pursuits, is key to reducing vulnerability and ensuring sustainable livelihoods. Policy support, including secure land tenure, climate adaptation strategies, and access to global markets, is essential to empower pastoralists in their efforts to adapt to a changing environment. Through support to pastoralists in these areas, governments, international organizations, and local communities can help ensure that pastoralism remains a viable and sustainable livelihood for millions of people in the Global South. In the face of climate change, the resilience of pastoralist communities will depend on their ability to harness both traditional knowledge and modern innovations, ensuring their continued relevance in the years to come.

References

Abdela, U. (2024). Assessment of community-driven drought risk management strategies in pastoral and agro-pastoral district of Bale zones south east Ethiopia. *Frontiers in Environmental Science, 12*, 1411138.

Aderinoye-Abdulwahab, S., & Ganiyu, A. R. (2025). Changing paradigms in addressing climate shocks among pastoralists. In *Exploring the unseen hazards of our world* (p. 19).

Agrawal, A., Erbaugh, J., & Pradhan, N. (2023). The commons. *Annual Review of Environment and Resources, 48*(1), 531–558.

Ahmed, M., Mohamed, M. D., & Muhammad, F. (2023). Pastoralists adaptation strategies, and resilience capacity to climate change in Somalia: A scoping review. https://doi.org/10.21203/rs.3.rs-2535309/v1

Akinyetun, T. S., & Ogunbodede, N. E. (2023). Conflict weather: Climate change as a driver of pastoralist conflicts in the Lake Chad Region. *Jurnal Politik, 9*(1), 2.

Aliyu, S. S., & Sama'ila, A. (2023). Pastoralist transhumance and conflicts in the Sahelian zone of the Nigeria-Niger borderlands. *South Asian Research Journal of Arts, Language and Literature, 5*(4), 165–172.

Azerigyik, R. A., Owusu-Ansah, J., Poku-Boansi, M., & Agyei, F. K. (2024). Contextualising sustainable transhumant pastoralism: A systematic review. *Sustainable Development*. https://doi.org/10.1002/sd.3273

Azimi, M., Sharifian, A., Riazinia, V., Arzani, H., & Abbaspour, K. C. (2022). Pastoral community's vulnerability under extreme floodings accelerated by rangeland degradation among Turkmen transhumant, Northern Iran. *Journal of Flood Risk Management, 15*(4), Article e12842.

Banerjee, R. R., Dhulipala, R., & Paliwal, A. (2023). Workshop report on digital and financial services in the livestock sector in India. Nairobi, Kenya: ILRI. https://cgspace.cgiar.org/server/api/core/bitstreams/f51afefb-dafb-4e23-ae48-3fb1a4b176de/content

Barrow, E., & Fellow, R. (2022). Pastoralists—The solution to sustainable dry landscape management, yet undermined and seen as the "problem". *Rights and Resources*.

Basupi, L. V. (2022). Pastoralism and livelihoods in the Global South. In *The Routledge handbook on livelihoods in the Global South* (pp. 302–311).

Behnke, R. H. (2021). Grazing into the Anthropocene or back to the future? *Frontiers in Sustainable Food Systems, 5*, Article 638806.

Birhanu, Z., Ambelu, A., Tesfaye, A., Berhanu, N., Kassahun, W., Daba, T., & Woldemichael, K. (2021). Prevalence of household food insecurity and associated factors in drought-prone pastoralist communities in Borana, Oromia, Ethiopia. *Ethiopian Journal of Health Development*, *35*(1).

Boas, I. (2022). The diversification of pastoralist herding: Navigating socio-climatic risk via mobile technologies. *Journal of Ethnic and Migration Studies, 48*(14), 3433–3449.

Campbell, T. (2022). Climate change policy narratives and pastoralism in Ethiopia: New concerns, old arguments? *Nomadic Peoples, 26*(1), 106–133.

Daum, T., Ravichandran, T., Kariuki, J., Chagunda, M., & Birner, R. (2022). Connected cows and cyber chickens? Stocktaking and case studies of digital livestock tools in Kenya and India. *Agricultural Systems, 196*, Article 103353.

Devereux, S. (2006). *Vulnerable livelihoods in Somali region, Ethiopia*. Brighton: Institute of Development Studies, 196p. https://opendata.uni-halle.de/bitstream/1981185920/106854/401/642906769.pdf

Duale, M. M. (2024). Resilience and transformation: Examining pastoral development initiatives in Ethiopia—Article review. *Open Access Library Journal, 11*(2), 1–17.

Dutta Gupta, T., Hassan, G. M., Abdi, A. N., Madurga Lopez, I. M., Liebig, T. I., Santa Cruz, L. M., & Pacillo, G. (2023). *How does climate exacerbate root causes of conflict in Kenya? Climate security pathway analysis*. FACT-SHEET 2023/1. CGIAR Focus Climate Security. https://cgspace.cgiar.org/bitstreams/11d679f2-ae01-4a2f-95e1-a89b298cdf02/download

Flintan, F. E., Robinson, L. W., & Allen, M. (2021). A review of tenure and governance in the pastoral lands of East and West Africa. https://cgspace.cgiar.org/server/api/core/bitstreams/9a090fc4-0e3b-48c2-9993-f085d45e11eb/content

Fonchingong Che, C., & Bang, H. N. (2024). Towards a "social justice ecosystem framework" for enhancing livelihoods and sustainability in pastoralist communities. *Societies, 14*(11), 239.

Fre, Z., Tsegay, B., Teka, A. M., Kenton, N., & Livingstone, J. (2022). *Social protection, pastoralism and resilience in Ethiopia*. Taylor Francis Limited.

Habte, M., Eshetu, M., Maryo, M., Andualem, D., & Legesse, A. (2022). Effects of climate variability on livestock productivity and pastoralists perception: The case of drought resilience in Southeastern Ethiopia. *Veterinary and Animal Science, 16*, Article 100240.

Herrera Calvo, P. M. (2024). Global land outlook: Thematic report on rangelands and pastoralists. United Nations Convention to Combat Desertification. https://cgspace.cgiar.org/bitstreams/2af249c9-abac-4274-8fec-0ba448c5d32d/download

Inman, E. N. (2024). "There are no true Himbas anymore": Exploring the dynamics of the Himba culture and land use in the face of change in Kunene Region, Namibia. *Sustainability, 16*(4), 1582.

IPC. (2024). Integrated food security phase classification. https://www.ipcinfo.org/

IPCC. (2023). Summary for policymakers. In Core Writing Team, H. Lee & J. Romero (Eds.), *Climate Change 2023: Synthesis Report. Contribution of Working Groups I, II and III to the Sixth Assessment Report of the Intergovernmental Panel on Climate Change* (pp. 1–34). https://doi.org/10.59327/IPCC/AR6-9789291691647.001. https://www.ipcc.ch/report/ar6/syr/downloads/report/IPCC_AR6_SYR_SPM.pdf

Kaguai, P. W. (2023). *Assessing the drought adaptation strategies and institutional interventions in the pastoral livelihood system in Narok County, Kenya* (Doctoral dissertation, University of Nairobi).

Kirui, L. K., Jensen, N. D., Obare, G. A., Kariuki, I. M., Chelanga, P. K., & Ikegami, M. (2022). Pastoral livelihood pathways transitions in northern Kenya: The process and impact of drought. *Pastoralism, 12*(1), 23.

LD4D. (2024). Livestock data for decisions: Digital system helps Ethiopia's pastoralist communities prepare for the unpredictable. https://livestockdata.org/member-spotlight/digital-system-helps-ethiopias-pastoralist-communities-prepare-unpredictable

Little, P. D. (2021). Does livelihood and asset diversification contribute to pastoralist resilience? The case of Il Chamus, Baringo county, Kenya, 1980–2018. *Nomadic Peoples, 25*(2), 181–205.

Lumborg, S., Tefera, S., Munslow, B., & Mor, S. M. (2021). Examining local perspectives on the influence of climate change on the health of Hamer pastoralists and their livestock in Ethiopia. *Pastoralism, 11*(1), 1–17.

Mabumbo, D., Nohayi, N., Sanchez Ramirez, J. C., Dessalegn, M., & Ruckstuhl, S. (2024). *Navigating climate resilience in fragile settings: A retrospective analysis of the 2023 flood impacts, early warning, and response in Ethiopia's Somali Region* (p. 30). International Water Management Institute (IWMI). CGIAR Initiative on Fragility, Conflict, and Migration. https://cgspace.cgiar.org/bitstreams/1f4c79fc-7092-4e84-8408-2ccfe8f85e81/download

Makokha, C., Jaquez, C., & Reid, E. (2022). Innovations for pastoralists and agro-pastoralists in fragile and conflict-affected settings.

Manzano, P., Burgas, D., Cadahía, L., Eronen, J. T., Fernández-Llamazares, Á., Bencherif, S., & Stenseth, N. C. (2021). Toward a holistic understanding of pastoralism. *One Earth, 4*(5), 651–665.

Mbih, R. A., Ndzeidze, S. K., Wanyama, D., & Mbuh, M. J. (2022). Challenges of transhumance in Northwest Cameroon. *SN Social Sciences, 2*(10), 208.

McGuirk, E. F., & Nunn, N. (2025). Transhumant pastoralism, climate change, and conflict in Africa. *Review of Economic Studies, 92*(1), 404–441.

McKay, S. (2023). Common-pool resources and democracy. *Critical Review of International Social and Political Philosophy*, 1–24.

Moreda, F., Kassa, G., & Tefera, S. (2024). Resilience in changing political and social environments: The Garri pastoralists in southern frontiers of Ethiopia. https://doi.org/10.52326/jss.utm.2024.7(2).08

Mutisya-Kioko, E. (2024). History of pastoralist strategies and resilience in Northern Kenya and Southern Ethiopia. *Oxford Research Encyclopedia of African History*. https://doi.org/10.1093/acrefore/9780190277734.013.1265

Naess, L. O., Wangari-Muneri, E., Nightingale, A. J., & Mehta, L. (2025). Climate change and the operation of power: Intersectionality, dispossession, and knowledge politics in pastoral communities. *The Journal of Peasant Studies*, 1–26.

Ndiritu, S. W. (2021). Drought responses and adaptation strategies to climate change by pastoralists in the semi-arid area, Laikipia County, Kenya. *Mitigation and Adaptation Strategies for Global Change, 26*, 1–18.

Neupane, M., Joshi, R., Bhandari, D., & Awasthi, N. (2022). Perception of transhumant herders on climate change and their adaptation strategies. *Our Nature, 20*(1), 1–13.

Ng'ang'a, T. W. (2022). *Socio-economic and technical assessment of adaptive capacity to climate change adaptation among the transitioning pastoralist and agro-pastoralist households of Laikipia County, Kenya* (Doctoral dissertation, University of Nairobi).

Njoki, F. W. (2023). *Environmental stress and conflicts in the Sahel region: A case study of Mali* (Doctoral dissertation, University of Nairobi). https://erepository.uonbi.ac.ke/bitstream/handle/11295/166578/Njoki%20Fridah%20Wanjiru%20-%20Project.pdf?sequence=1&isAllowed=y

Nori, M., & Scoones, I. (2023). Rethinking policies for pastoralists–governing the rangelands. *The Rangeland Journal, 45*(2), 53–66.

Ntumva, M. E. (2022a). Farmer-pastoralist conflicts management approaches in Sub-Saharan Africa: Insights into their strengths and pitfalls. *Journal of African Studies and Development, 14*(1), 12–26.

Ntumva, M. E. (2022b). Land conflict dynamics in Africa: A critical review on farmer-pastoralist conflict perspectives. *International Journal of Peace and Development Studies, 13*(1), 17–28.

Nulkar, G. (2024). Traditional knowledge and sustainable livelihoods. In *The economics of sustainable development: A machine-generated literature overview* (pp. 129–176). Springer Nature Singapore.

Omondi, I., Baltenweck, I., Kinuthia, E., Kirui, L., Njoroge-Wamwere, G., Bett, B., & Kiara, H. (2021). Mobile veterinary clinics in the drylands of Kenya: Securing pastoralists' livelihoods by bringing services close. *Development in Practice, 31*(5), 561–579.

Otieno, S., Abade, E., & Oludhe, C. (2024). A comprehensive review of climate information systems in the arid and semi-arid lands (ASALs) of Kenya for enhanced decision-making process. *Information Development*, 02666669241287498.

Ouédraogo, K., Zaré, A., Korbéogo, G., Ouédraogo, O., & Linstädter, A. (2021). Resilience strategies of West African pastoralists in response to scarce forage resources. *Pastoralism, 11*(1), 16.

Ouko, O. C. (2024). Participatory assessment of climate and disaster risk among pastoral communities of Africa—A case study of Kajiado County, Kenya. *African Journal of Emerging Issues, 6*(12), 53–76.

Parlasca, M. C. (2021). A vital technology: Review of the literature on mobile phone use among pastoralists. *Journal of International Development, 33*(4), 780–799.

Rajput, P. (2024). Sustainable pastoralism amid changing climate: Eco-centric practices of Bakarwal nomads in the Western Himalayas. *GeoJournal, 90*(1), 14.

Rigby, J. M., Stawarz, K., Preist, C., Saeed, A., Stokes, K., Elmi, M., & Michaelides, K. (2023). Exploring the information needs of Somaliland pastoralists: Design considerations for digital climate adaptation services. In *Proceedings of the 2023 ACM Designing Interactive Systems Conference* (pp. 1548–1565).

Robinson, L., & Flintan, F. (2022). Can formalisation of pastoral land tenure overcome its paradoxes? Reflections from East Africa. *Pastoralism, 12*(1), 34.

Sani Ibrahim, S., Ozdeser, H., Cavusoglu, B., & Abdullahi Shagali, A. (2021). Rural migration and relative deprivation in agro-pastoral communities under the threat of cattle rustling in Nigeria. *SAGE Open, 11*(1), 2158244020988856.

Scoones, I. (2021). Pastoralists and peasants: Perspectives on agrarian change. *The Journal of Peasant Studies, 48*(1), 1–47.

Senda, T. S., Robinson, L. W., Gachene, C. K., & Kironchi, G. (2022). Formalization of communal land tenure and expectations for pastoralist livelihoods. *Land Use Policy, 114*, Article 105961.

Sewando, P. T. (2022). Efficacy of risk reducing diversification portfolio strategies among agro-pastoralists in semi-arid area: A modern portfolio theory approach. *Journal of Agriculture and Food Research, 7*, Article 100262.

Sharifian, A., Fernández-Llamazares, Á., Wario, H. T., Molnár, Z., & Cabeza, M. (2022). Dynamics of pastoral traditional ecological knowledge: A global state-of-the-art review. *Ecology and Society, 27*(1), 14.

Singh, R., Bhutia, K. S., Bhutia, T. U., & Babu, S. (2022). Rangeland conservation, pastoralist displacement, and long-term implications of a grazing ban in the Indian Himalaya. *Ecology, Economy and Society—the INSEE Journal, 1*(1), 195–221.

Slayi, M., Zhou, L., Thamaga, K. H., & Nyambo, P. (2024). The role of social inclusion in restoring communal rangelands in Southern Africa: A systematic review of approaches, challenges, and outcomes. *Land, 13*(9), 1521.

Stavi, I., Roque de Pinho, J., Paschalidou, A. K., Adamo, S. B., Galvin, K., de Sherbinin, A., & van der Geest, K. (2022). Food security among dryland pastoralists and agropastoralists: The climate, land-use change, and population dynamics nexus. *The Anthropocene Review, 9*(3), 299–323.

Sulieman, H. M., & Young, H. (2023). The resilience and adaptation of pastoralist livestock mobility in a protracted conflict setting: West Darfur, Sudan. *Nomadic Peoples, 27*(1), 3–31.

Tesfaye, B. (2022). *Climate change and conflict in the Sahel*. Council on Foreign Relations. Discussion Paper Series on Managing Global Disorder No. 11 November 2022. https://cdn.cfr.org/sites/default/files/report_pdf/Climate%20Change%20and%20Conflict%20in%20the%20Sahel.pdf

Thornton, P. K., van de Steeg, J., Notenbaert, A., & Herrero, M. (2009). The impacts of climate change on livestock and livestock systems in developing countries: A review of what we know and what we need to know. *Agricultural systems, 101*(3), 113–127.

Timpong-Jones, E. C., Samuels, I., Sarkwa, F. O., Oppong-Anane, K., Majekodumni, A. O., & Wario, H. T. (2023). Transhumance pastoralism in West Africa—Its importance, policies and challenges. *African Journal of Range & Forage Science, 40*(1), 114–128.

Tofu, D. A. (2024). Evaluating the impacts of climate-induced east Africa's recent disastrous drought on the pastoral livelihoods. *Scientific African, 24*, Article e02219.

Tofu, D. A., Dilbato, T., Fana, C., Dirbaba, N. B., & Tesso, G. (2025). Analysis of vulnerability, its drivers, and strategies applied towards reducing the pastoral and agro-pastoral livelihood vulnerability to climatic shocks. *Scientific Reports, 15*(1), 2567.

Tofu, D. A., Fana, C., Dilbato, T., Dirbaba, N. B., & Tesso, G. (2023). Pastoralists' and agro-pastoralists' livelihood resilience to climate change-induced risks in the Borana zone, south Ethiopia: Using resilience index measurement approach. *Pastoralism, 13*(1), 1–14.

Tugjamba, N., Walkerden, G., & Miller, F. (2023). Adapting nomadic pastoralism to climate change. *Climatic Change, 176*(4), 28.

Turnbull, S., & Harrison, C. (2024). Shepherd's eye in the sky: The potential for Afriscout digital grazing maps to improve pastoralists' grazing and migration decisions.

Tuwei, D., & Tully, M. (2021). The role of change agents in the adaptation and use of mobile money services in Kenya. *Journal of African Media Studies, 13*(1), 89–102.

Van Klingeren, F., & de Graaf, N. D. (2021). Heterogeneity, trust and common-pool resource management. *Journal of Environmental Studies and Sciences, 11*, 37–64.

Xiandong, L. (2023). A study of livelihood transformation strategies for digitally driven pastoralists' ecological perceptions. https://doi.org/10.21203/rs.3.rs-3680718/v1

Xie, B., Wuliji, M., & Zhao, L. (2024). Integrating traditional ecological knowledge into land use and land cover change assessments, pastoralist communities in northwest inner Mongolia China. *Land, 13*(12), 1979.

Yator, M. (2024). Adaptation measures adopted by pastoralist livestock farmers in Kenya in response to climate change. *Journal of Aquatic and Terrestrial Ecosystems, 2*(2), 22–36.

Zhang, Q. (2023). Land use and adaptive governance under climate change: Analysis of four cases in pastoral areas of China. *Frontiers in Environmental Science, 11*, Article 922417.

CHAPTER 5

Livelihood Resilience in Agriculture-Dependent Communities in the Global South in the Face of Climate Change

Abstract Climate change poses significant challenges to agriculture-dependent communities in the Global South, where recurrent droughts, floods, and shifting weather patterns threaten food security and agricultural productivity. These environmental shifts compel the adoption of adaptive agricultural practices, such as crop diversification, agroforestry, conservation agriculture, and the use of climate-resilient crop varieties, which have proven successful in enhancing resilience. However, access to resources and knowledge remains a critical barrier; farmers in these regions often lack adequate education, extension services, and timely information on climate adaptation techniques. This impedes their ability to fully capitalize on sustainable agricultural practices. Case studies highlight both success stories and ongoing challenges. Successful adaptation projects demonstrate the importance of localized approaches, but obstacles such as limited access to markets, technology, and financial resources persist. Policy interventions are crucial to supporting agricultural resilience. National and international policies must focus on improving finance, infrastructure, and rural development to foster adaptive capacity and reduce vulnerability. Effective policy frameworks that integrate climate change adaptation into agricultural systems are essential for long-term sustainability. This chapter therefore emphasizes on the

N. P. Awazi, *Building Resilience*, Palgrave Studies in Climate Resilient Societies, https://doi.org/10.1007/978-3-031-93791-0_5

complex relationship between climate change, agriculture, and community resilience in the Global South, offering insights into practical solutions and identifying areas where further support is needed.

Keywords Climate change · Agriculture · Resilience · Adaptation · Global South · Policies · Resources

5.1 Climate Change and Agriculture in the Global South

Climate change impacts are disproportionately felt in the Global South, where the majority of developing countries are located (Ngcamu, 2023). In these regions, agriculture is the backbone of economies, providing livelihoods for millions of people and ensuring food security for the population (Asma et al., 2023). However, climate change is intensifying environmental stressors, such as recurrent droughts, floods, and shifting weather patterns, which are significantly affecting agricultural production and food security (Table 5.1). Agriculture in the Global South is highly vulnerable to climate change due to several factors (Raj et al., 2022). Firstly, many countries in this region rely heavily on rain-fed agriculture, which makes them sensitive to changes in precipitation patterns. For instance, in Sub-Saharan Africa, about 95% of farming is rain-fed (Lamptey, 2022; Msongaleli, 2024). Secondly, most agricultural systems are based on traditional practices that may not be equipped to adapt to rapidly changing climatic conditions (Mbah et al., 2021). Thirdly, the lack of financial and technological resources further limits the ability of farmers to implement climate-resilient agricultural practices (Samuel et al., 2021).

Droughts are among the most common and devastating climatic events experienced in the Global South (Lottering et al., 2021). In regions such as East Africa, recurrent droughts have led to severe water shortages, crop failures, and livestock deaths. According to a report by the United Nations Office for Disaster Risk Reduction (UNDRR), droughts in the Horn of Africa have caused food insecurity affecting over 16 million people, with countries like Somalia, Ethiopia, and Kenya being hit hardest (UNDRR, 2024). Droughts have a direct impact on crop yields, especially staple crops like maize, sorghum, and millet, which are crucial for food security in these areas. One of the key challenges of droughts is

Table 5.1 Climate change impacts on agriculture in the Global South

Climate change impact	*Description*	*Key examples*	*Key references*
Droughts	Increased frequency and severity of droughts leading to water shortages, crop failures, and livestock deaths, especially in regions reliant on rain-fed agriculture	- **Horn of Africa**: Droughts affecting over 16 million people in Somalia, Ethiopia, and Kenya - **Southern Africa (2016–2017)**: Maize production fell by over 30% due to drought	Lottering et al. (2021), UNDRR, (2024), IFPRI (2024), and Panda et al. (2022)
Floods	More frequent and severe flooding that damages crops, infrastructure, and communities, often leading to salinization of soil and loss of valuable topsoil	- **Southeast Asia**: Over 20 million people affected annually by floods in countries like Bangladesh, India, and the Philippines - Floods leading to rice and maize crop losses in Bangladesh and Vietnam	Montesclaros and Teng (2021), Waqas et al. (2024), Yuen et al. (2022), and Ali (2023)
Shifting weather patterns	Alteration of traditional weather patterns, affecting planting and harvesting times, reducing crop yields, and increasing the risk of pests and crop diseases	- **West Africa (Sahel)**: Delayed rainy seasons reduce crop yields by 30% - **Southern Africa**: Maize yields decreased by 20–30% due to temperature increases	Yuan et al. (2024), Porkka et al. (2021), Saley and Salack (2023), and Siatwiinda et al. (2021)
Crop failures and yield reduction	Decreased agricultural productivity due to droughts, floods, and shifting weather patterns, particularly for staple crops like maize, sorghum, millet, and rice	- **Southern Africa**: Maize production drop in Zimbabwe, Zambia, and Malawi during the 2016–2017 drought - **India**: Decreased groundwater levels impacting irrigation and crop productivity	Lamptey (2022), Msongaleli (2024), and Mbah et al. (2021)
Water scarcity	Decreased availability of water resources, both for irrigation and livestock, compounded by overexploitation of groundwater during drought conditions	- **India**: Overexploitation of groundwater resources during droughts, reducing water tables - **Sub-Saharan Africa**: Rain-fed agriculture faces increased vulnerability due to water scarcity	Panda et al. (2022) and Raj et al. (2022)

(continued)

Table 5.1 (continued)

Climate change impact	*Description*	*Key examples*	*Key references*
Food security impacts	Climate change directly affects food availability, access, and utilization, increasing food insecurity and malnutrition, especially in rural areas	- **Horn of Africa**: 16 million people affected by food insecurity due to drought - **Global South**: Increased poverty and malnutrition due to reduced agricultural productivity and food price hikes	WHO (2021), Ambikapathi et al. (2022), and World Bank (2020)
Increased malnutrition	Climate change exacerbates malnutrition due to reduced agricultural diversity, higher food prices, and increased vulnerability of rural populations to food insecurity	- **Sub-Saharan Africa**: Reduced availability of diverse crops increases malnutrition in rural children - **South Asia**: Changing crop patterns lead to nutritional deficiencies	WHO (2021) and Ambikapathi et al. (2022)
Economic vulnerability	Climate change impacts undermine economic stability in agricultural sectors, pushing more people into poverty due to reduced crop production and rising food prices	- World Bank (2020): Estimated that 100 million people could fall into extreme poverty by 2030 due to agriculture disruptions - **Smallholder farmers**: Vulnerable to economic hardship from reduced productivity and increased costs	World Bank (2020)

that they not only reduce the availability of water for irrigation but also significantly lower soil moisture levels, rendering the land less fertile. This leads to decreased agricultural productivity and often results in crop failures. The 2016–2017 drought in Southern Africa, for instance, led to maize production falling by over 30% in Zimbabwe, Zambia, and Malawi, affecting food security in these countries (IFPRI, 2024). Moreover, droughts lead to the depletion of groundwater resources, which are essential for irrigation, thereby compounding the problem. In countries like India, where groundwater is a critical resource for agriculture, overexploitation during drought conditions has led to a decrease in water tables, further exacerbating water scarcity (Panda et al., 2022).

Floods are another climate-related hazard that has become more frequent and severe in many parts of the Global South. Unlike droughts, floods can cause immediate and widespread damage to crops, infrastructure, and communities. In countries like Bangladesh, India, and the Philippines, floods have caused widespread devastation, displacing millions and damaging agricultural lands. Recurrent floods in Southeast Asia, for example, affect more than 20 million people annually, damaging over 4 million hectares of agricultural land, leading to crop losses of rice, maize, and vegetables (Montesclaros & Teng, 2021; Waqas et al., 2024). In such instances, farmers lose not only their crops but also valuable soil nutrients, as floodwaters often wash away topsoil, which is vital for crop growth. In addition, the standing water increases the risk of waterborne diseases, which further complicates recovery efforts. Flooding also poses long-term challenges to agricultural productivity. Floodwaters often leave behind salinized soil, especially in coastal regions, which makes it difficult to grow certain crops. The inundation of rice fields in countries like Bangladesh and Vietnam has resulted in the loss of crucial rice-growing land and threatens food production (Ali, 2023; Yuen et al., 2022). The salinization of agricultural soils is a growing concern, particularly as sea levels rise due to climate change, further exacerbating the vulnerability of the agricultural sector.

Another significant impact of climate change in the Global South is the alteration of traditional weather patterns, which has a profound effect on growing seasons and agricultural production (Yuan et al., 2024). Many farmers in the Global South rely on specific seasonal weather patterns to determine planting and harvesting times. However, as climate change leads to unpredictable and shifting weather patterns, these schedules are becoming increasingly unreliable. In regions like West Africa, where the timing of the rainy season is critical for the cultivation of crops like maize, millet, and groundnuts, a delay in the onset or end of the rains can result in crop failures. According to studies by Porkka et al. (2021) and Saley and Salack (2023), the delayed onset of rains in the Sahel has resulted in a 30% reduction in crop yields over the past three decades. The shifting of seasonal weather patterns is not only reducing productivity but also increasing the risk of pest infestations and crop diseases, as warmer temperatures and unpredictable rainfall create favorable conditions for pests like locusts. In addition, the warming of the atmosphere and changes in rainfall have been linked to altered crop suitability. For example, in southern Africa, changing weather patterns are affecting the cultivation of crops like maize, which are sensitive to temperature changes. A rise in average temperatures by 2 °C could lead to a 20–30% decrease in maize yields in regions like Zimbabwe and Zambia (Siatwiinda et al., 2021). This reduction in productivity exacerbates food insecurity, as maize is a staple food in many parts of Africa.

The compounded impacts of recurrent droughts, floods, and shifting weather patterns have direct implications for food security in the Global South. Food security is defined as the availability, access, and utilization of food, and climate change impacts each of these dimensions in various ways. First, the reduction in crop yields due to droughts and floods limits the availability of food. Second, the economic hardship faced by farmers due to these climatic shocks makes it harder for them to access food, as incomes are diminished and markets are disrupted. The World Bank (2020) estimates that climate change could push an additional 100 million people into extreme poverty by 2030, with agriculture being one of the most affected sectors. The economic losses resulting from agricultural disruptions are felt across the entire food value chain, from production to consumption. In many developing countries, agricultural households are also net buyers of food, meaning that price increases due to reduced supply exacerbate food insecurity. In addition to the direct impact on food availability, climate change also exacerbates malnutrition in vulnerable populations. The reduction in the diversity of available foods due to changing agricultural production patterns, coupled with the increasing cost of food, leads to poorer diets and greater nutritional deficiencies (Ambikapathi et al., 2022; WHO, 2021). Children in rural areas are particularly affected by these shifts, as they face higher rates of stunting and malnutrition due to a lack of access to adequate nutrition.

Despite the challenges posed by climate change, there are numerous adaptation strategies that can help mitigate the impacts on agriculture in the Global South. One approach is the adoption of climate-resilient agricultural practices, such as drought-resistant crop varieties, agroforestry, and improved water management techniques (Erezi et al., 2023; Hussein, 2024). The use of early warning systems and climate forecasting can also help farmers plan their activities around anticipated weather patterns.

In addition, governments and international organizations must prioritize investments in agricultural research and infrastructure. Financial support for smallholder farmers, who make up the majority of agricultural producers in the Global South, is crucial in helping them adopt climate-smart technologies and practices (Villalba et al., 2024). Equally, regional cooperation and knowledge-sharing can help spread successful adaptation strategies across borders.

The impacts of recurrent droughts, floods, and shifting weather patterns on agriculture in the Global South are therefore severe and far-reaching. Climate change is exacerbating existing vulnerabilities, undermining agricultural productivity, and threatening food security. As the Global South continues to experience the consequences of a changing climate, it is essential to implement adaptation and mitigation strategies that promote climate-resilient agricultural systems. These efforts, coupled with international support and investment in climate-smart agriculture, can help to build a more sustainable and food-secure future for the region.

5.2 Adaptive Agricultural Practices in the Global South in the Face of Climate Change

Climate change presents one of the most significant challenges for agriculture globally, but its effects are particularly pronounced in the Global South, where many countries are heavily dependent on agriculture for livelihoods and economic stability. The region's agriculture is often vulnerable to changing weather patterns, extreme events like droughts and floods, and shifting growing seasons. In response to these challenges, adaptive agricultural practices are essential to building resilience. These practices not only help mitigate the impacts of climate change but also provide opportunities for sustainable development, enhancing food security and economic stability. The variety of adaptive agricultural practices implemented across the Global South, include crop diversification, conservation agriculture, agroecology, agroforestry, and climate-resilient crop varieties (Table 5.2).

Crop diversification is one of the most widely recommended adaptive strategies in the face of climate change (Gebre et al., 2023; Mohammed et al., 2021). This practice involves growing a variety of crops rather than a single staple crop, which reduces the risk of crop failure due to

Table 5.2 Adaptive agricultural practices in the Global South in the face of climate change

Adaptive agricultural practice	*Description*	*Key examples*	*Key references*
Crop diversification	Growing a variety of crops to reduce the risk of crop failure due to changing weather patterns. Includes both traditional and resilient crop varieties	- **Ethiopia (Highlands)**: Diversification with drought-resistant crops like millet and sorghum alongside traditional grains like teff - **West Africa**: Planting drought-tolerant crops with maize to ensure food stability during poor rainfall years	Mohammed et al. (2021), Gebre et al. (2023), Makate et al. (2022), Bekuma et al. (2024), Turner and Ayantunde (2022), and Zakari et al. (2022)
Conservation Agriculture (CA)	Focus on improving soil health, conserving water, and reducing dependency on chemical fertilizers. Involves minimal soil disturbance, crop rotation, and cover crops	- **Zambia**: Adoption of minimum tillage to reduce soil erosion and retain moisture, improving resilience to drought - **Kenya and Malawi**: Use of cover crops like legumes to improve soil fertility and water retention	Tembo et al. (2025) and Marambanyika (2022)
Agroecology	Integrates ecological principles into farming, emphasizing biodiversity, local knowledge, and organic practices for resilient farming systems	- **Brazil**: Movimento dos Trabalhadores Rurais Sem Terra (MST) promotes organic farming and crop rotation to increase biodiversity and soil health - **Various regions**: Use of polyculture, local seeds, and organic methods	Milhomens et al. (2021) and Stratton et al. (2021)

(continued)

Table 5.2 (continued)

Adaptive agricultural practice	*Description*	*Key examples*	*Key references*
Agroforestry	Combining trees with crops or livestock for benefits like carbon sequestration, soil fertility, and improved water retention, alongside income diversification	- **India (Punjab)**: Integration of trees like acacia and eucalyptus with wheat and rice crops to improve soil quality and water retention - **Kenya**: Integration of trees like Grevillea with crops like maize for improved resilience	Ministry of Agriculture and Farmers Welfare (2018), Verma et al. (2021), Jinger et al. (2022), and ICRAF (2015)
Climate-resilient crop varieties	Development and adoption of crops bred for resilience to extreme conditions such as drought, heat, and floods, maintaining productivity despite climate stress	- **Sub-Saharan Africa**: Introduction of drought-resistant maize varieties increasing productivity by 30% - **India**: Salt-tolerant rice varieties like "Swarna Sub1" enable rice cultivation in saline environments	AATF (2024), Wanjira (2021), Mpala and Simatele (2024), and Nayak et al. (2022)
Water harvesting techniques	Methods to capture and store water during rainy periods for use during droughts, improving water availability in arid regions	- **Sahel Region (West Africa)**: Techniques like water catchment and construction of small reservoirs to store water during the rainy season for use in droughts	Sanogo et al. (2024)

(continued)

Table 5.2 (continued)

Adaptive agricultural practice	*Description*	*Key examples*	*Key references*
Improved irrigation systems	Adoption of more efficient irrigation methods to conserve water, including solar-powered pumps	- **India and Nepal**: Use of solar-powered irrigation pumps to improve water-use efficiency and reduce reliance on fossil fuels	Neupane et al. (2022)
Sustainable livestock management	Adaptive practices that focus on improving livestock resilience to climate change, including better grazing practices and feed management	- **Various regions in the Global South**: Techniques for managing livestock health and improving productivity in the face of shifting climate patterns, including better grazing rotations and feed practices	(General Reference to Global Adaptation Practices)

unpredictable weather patterns. Diversification can include both traditional crops and new varieties that are more resilient to changing climatic conditions. One prominent example of crop diversification is in the highlands of Ethiopia, where farmers have diversified their crops to include both traditional grains like teff and new, more drought-resistant varieties such as millet and sorghum. According to studies by Makate et al. (2022) and Bekuma (2024), this diversification has led to improved resilience in the face of erratic rainfall patterns. Similarly, in regions of West Africa, crop diversification has been implemented to reduce the vulnerability of subsistence farmers who rely heavily on crops like maize, which are particularly susceptible to changing rainfall and temperature patterns (Turner & Ayantunde, 2022; Zakari et al., 2022). Through the planting of drought-tolerant crops alongside maize, farmers are able to ensure a more stable food supply even in years of poor rainfall.

Conservation agriculture (CA) is another sustainable approach which involves practices that improve soil health, conserve water, and reduce the dependency on external inputs like chemical fertilizers. It emphasizes three main principles: minimal soil disturbance, crop rotation, and the use of cover crops. CA is gaining traction as a climate adaptation strategy because it can increase soil fertility, improve water retention, and enhance resilience to extreme weather events such as floods and droughts. A key example of CA can be seen in Zambia, where farmers have adopted minimum tillage, which reduces soil erosion and helps retain moisture in the soil, making crops more resilient to drought conditions. A study by Tembo et al. (2025) found that farmers practicing CA saw improved crop yields even in seasons with reduced rainfall. Similarly, in countries like Kenya and Malawi, the practice of planting cover crops such as legumes has helped improve soil health, increase nitrogen levels, and reduce weed growth, leading to better crop performance despite climate variability (Marambanyika, 2022).

Agroecology equally offers a holistic approach to agriculture that integrates ecological principles into farming practices. It emphasizes biodiversity, ecosystem services, and the use of local knowledge to create sustainable and resilient farming systems. Agroecology encourages practices such as polyculture, organic farming, and the use of local seeds and varieties, which can increase resilience to climate change while promoting food sovereignty and reducing dependency on external inputs. In Brazil, the Movimento dos Trabalhadores Rurais Sem Terra (MST), or Landless Workers' Movement, has been a proponent of agroecology for decades. Their approach includes the use of organic farming techniques and crop rotation to increase biodiversity and soil health. According to studies by Milhomens et al. (2021) and Stratton et al. (2021), agroecological systems in Brazil have shown increased resilience to climate change, particularly in areas where conventional agriculture has struggled with soil degradation and water scarcity. Through the promotion of practices that enhance the natural resilience of ecosystems, agroecology has become a powerful tool for climate adaptation, particularly in regions where smallholder farmers are vulnerable to climate shocks.

Agroforestry is another promising adaptive practice that combines trees with agricultural crops or livestock. This practice is particularly important in the context of climate change, as trees provide a range of benefits, including carbon sequestration, soil fertility enhancement, and improved water retention. Agroforestry systems also offer diversification of income sources through the sale of timber, fruit, and other tree products. In India, the government's National Mission for Sustainable Agriculture (NMSA) has promoted agroforestry as a key climate adaptation strategy (Ministry of Agriculture and Farmers Welfare, 2018). Farmers in states like Punjab have been encouraged to integrate trees such as acacia and eucalyptus with their wheat and rice crops. These agroforestry systems have been found to improve soil quality, reduce the need for synthetic fertilizers, and provide shade for crops during extreme heat events, thus increasing overall productivity. Studies by Verma et al. (2021) and Jinger et al. (2022) highlight that agroforestry has led to better water management in semi-arid regions of India, where water scarcity is exacerbated by climate change. Similarly, in Kenya, the promotion of agroforestry through organizations like the World Agroforestry Centre (ICRAF) has helped farmers adapt to changing rainfall patterns (ICRAF, 2015). The integration of trees/shrubs like Grevillea, fruit trees and Calliandra with food crops such as maize and beans has improved soil fertility, provided fodder for livestock, and created more resilient farming systems. Farmers have reported increased yields and reduced vulnerability to droughts, particularly in arid and semi-arid regions.

One of the most direct ways to address the challenges posed by climate change is through the development and adoption of climate-resilient crop varieties. These varieties are bred to withstand extreme weather conditions, including drought, heat, and floods, while maintaining productivity. In Sub-Saharan Africa, the development of drought-resistant maize has been a key adaptation strategy. The African Agricultural Technology Foundation (AATF) and other organizations have worked on breeding maize varieties that are more tolerant of dry conditions (AATF, 2024). In countries like Kenya and Zimbabwe, the introduction of drought-tolerant maize has resulted in higher yields during periods of reduced rainfall, which is crucial for food security. According to a study by Wanjira (2021) and Mpala and Simatele (2024), farmers who adopted drought-resistant maize varieties experienced an increase in productivity by up to 30% compared to those planting conventional varieties. Similarly, in Asia, the development of salt-tolerant rice varieties has been crucial for coastal farmers who face the threat of salinization due to rising sea levels and increased flooding. In India, the introduction of salt-tolerant rice varieties like "Swarna Sub1" has allowed farmers to continue cultivating rice in saline environments, significantly improving food security and income (Nayak et al., 2022).

Apart from the strategies mentioned above, there are several other adaptive agricultural practices that are being implemented in the Global South. These include water harvesting techniques, the use of improved irrigation systems, and the promotion of sustainable livestock management practices. For example, in the Sahel region of West Africa, farmers have adopted techniques like water catchment and the construction of small reservoirs to store water during the rainy season, which can be used during periods of drought (Sanogo et al., 2024). Moreover, the practice of shifting to low-carbon farming technologies, such as the use of solar-powered irrigation pumps, has helped farmers in countries like India and Nepal improve water-use efficiency and reduce dependence on fossil fuels (Neupane et al., 2022). These practices, combined with improved management of natural resources, offer a comprehensive approach to tackling the challenges posed by climate change.

In conclusion, adaptive agricultural practices in the Global South, such as crop diversification, conservation agriculture, agroecology, agroforestry, and the development of climate-resilient crop varieties, have proven to be effective tools in addressing the challenges posed by climate change. By integrating these practices into farming systems, farmers in the Global South can build resilience to climate variability, enhance food security, and promote sustainable development. Continued support for these practices, through policies, research, and education, is essential for ensuring that agriculture in the Global South can withstand the increasingly unpredictable climate patterns of the future. As climate change continues to affect agricultural productivity, it is vital that adaptive practices are further expanded and tailored to the specific needs of local communities. By fostering innovation, collaboration, and sustainability, the Global South can chart a path toward climate-resilient agriculture that benefits both people and the planet.

5.3 Access to Resources and Knowledge in the Agricultural Sector in the Face of Climate Change in the Global South

The agricultural sector in the Global South faces multifaceted challenges that threaten food security, livelihoods, and economic development. Climate change, with its rising temperatures, erratic weather patterns, and extreme events such as droughts and floods, exacerbates these challenges. Smallholder farmers, who constitute a significant portion of the population in these regions, are especially vulnerable. Access to resources and knowledge is a critical factor in building resilience and enabling farmers to adapt to these challenges. In this context, education, extension services, and access to information play pivotal roles in improving agricultural practices, enhancing productivity, and promoting sustainable development.

Education in the agricultural sector is fundamental to equipping farmers with the necessary skills and knowledge to address climate change and improve agricultural productivity. It enables farmers to understand the changing climate patterns, adopt new technologies, and implement sustainable practices that can mitigate climate-related risks. Education empowers farmers to move away from traditional farming methods and embrace innovative practices that enhance food security and environmental sustainability (Adefila et al., 2024; Pandey et al., 2024). Formal education and vocational training programs tailored to the agricultural context can introduce farmers to the scientific basis of climate change and its implications. For instance, understanding how temperature fluctuations, altered rainfall patterns, and increased pests and diseases affect crop production can guide farmers in adopting climate-resilient practices. Furthermore, educating farmers on soil management, water conservation, and sustainable pest control methods can contribute to long-term resilience. In regions such as Sub-Saharan Africa and South Asia, where agriculture is the primary livelihood for the majority of the population, education has been shown to improve agricultural productivity and income generation. For example, a study by Wens et al. (2022) found that farmers who received training on modern agricultural practices achieved higher yields and were better able to cope with the effects of droughts and floods. Moreover, education fosters innovation, with farmers adopting new technologies such as drought-resistant crop varieties and climate-smart irrigation systems, which are crucial for adapting to climate change.

Extension services are another critical component in building resilience within agricultural communities. These services bridge the gap between research institutions and farmers, ensuring that farmers have access to up-to-date information on agricultural practices, climate change adaptation strategies, and government policies. Extension agents provide tailored advice and hands-on support to farmers, facilitating the adoption of new practices that can enhance productivity and mitigate climate risks. In many parts of the Global South, extension services are particularly important because they provide localized knowledge that is more relevant to the specific conditions and challenges faced by farmers. This includes advice on crop rotation, agroforestry, soil fertility management, and water harvesting techniques. According to a study by Antwi-Agyei and Stringer (2021), effective extension services can significantly improve agricultural outcomes by helping farmers understand local climate patterns and implement practices that enhance their resilience to climate variability. However, the effectiveness of extension services often depends on the availability of trained personnel, adequate funding, and the ability to reach remote or marginalized communities. In many countries in the Global South, extension services are underfunded and understaffed, limiting their reach and effectiveness. Furthermore, extension agents must be well-versed in climate change adaptation techniques to provide accurate and relevant advice. This necessitates continuous training and capacity-building for extension workers, ensuring that they remain up-to-date with the latest research and technologies in climate-resilient agriculture. In recent years, there has been a shift toward digital extension services, which leverage technology to disseminate information to farmers. This includes the use of mobile phones, internet-based platforms, and radio programs to reach a wider audience. Mobile phone-based platforms, such as those used in Kenya (e.g., the M-Farm platform), have proven to be effective in providing farmers with real-time market information, weather forecasts, and agricultural tips (Krell, 2021; Vrielink, 2023). These digital tools enable farmers to make informed decisions about planting, harvesting, and managing their crops, thereby improving their resilience to climate change.

Access to information is crucial in enabling farmers to adapt to climate change and make informed decisions. Information about climate forecasts, weather patterns, pest outbreaks, and market prices can help farmers plan their activities, reduce risks, and improve productivity. However, the accessibility of such information is often limited in rural areas of the Global South due to poor infrastructure, lack of literacy, and limited access to technology. The role of information systems in building resilience has been highlighted by studies such as Sultana and Luetz (2022) and Balogun et al. (2022), who argue that accurate, timely, and locally relevant information is essential for effective decision-making. For example, early warning systems that provide weather forecasts and information about climate-related risks can help farmers prepare for adverse events such as floods, droughts, or heatwaves. Such systems can be particularly useful in countries where climate variability is high, and farmers have limited access to traditional sources of information. In addition to weather-related information, market information is also crucial for farmers to make informed decisions about the timing of planting, harvesting, and selling their crops. By knowing the prevailing market prices, farmers can better plan their production and minimize post-harvest losses. This is particularly important for smallholder farmers, who often lack access to formal market channels and face exploitation by middlemen. Information about government policies and subsidies also plays a critical role in helping farmers access financial support for climate adaptation. In many cases, however, farmers in the Global South face challenges in accessing reliable information. Poor internet connectivity, low literacy rates, and the digital divide are significant barriers to accessing online platforms and information services. To overcome these challenges, alternative channels of information dissemination, such as community radio stations and village meetings, can be used to reach farmers. A study by Izuchukwu et al. (2023) emphasized the importance of community-based information networks in Africa, where farmers rely on local knowledge and peer-to-peer networks to exchange information about climate adaptation and agricultural practices.

5.4 Success Stories and Challenges of Agricultural Sector Adaptation to Climate Change in the Global South

The agricultural sector in the Global South faces significant challenges due to climate change, which has impacted crop yields, farming systems, and the livelihoods of millions of people. However, numerous adaptation strategies have been developed to mitigate these challenges, showing resilience and the potential for agricultural systems to adjust to the changing climate. Some success stories include climate-smart agriculture in Kenya, Agroecology in Brazil, the Zero Budget Natural Farming model in India, and sustainable irrigation in Egypt.

Kenya, a country heavily reliant on agriculture, has experienced significant climate variability in recent years, with prolonged droughts and floods damaging crops (Kogo et al., 2021). In response, the Kenyan government, along with local farmers and international organizations, have adopted Climate-Smart Agriculture (CSA) practices. These include practices such as conservation tillage, the use of drought-resistant seed varieties, water harvesting techniques, and agroforestry. A notable success story comes from the village of Makueni in Eastern Kenya, where a combination of CSA practices has improved agricultural productivity while conserving resources. Farmers in Makueni were introduced to drought-resistant maize varieties and small-scale rainwater harvesting techniques. By building water storage systems such as ponds and improving soil fertility through organic composting, farmers have been able to increase food production even during periods of drought (Kalele et al., 2021; Odhiambo et al., 2021). These changes have increased crop yields, enhanced food security, and boosted farmers' income, demonstrating how adaptation to climate change can be achieved through innovative agricultural practices.

Agroecology has emerged as a key strategy for sustainable farming in Brazil, particularly in the semi-arid Northeast region, which is prone to frequent droughts and desertification. The Brazilian government, alongside various NGOs and local communities, has promoted agroecological practices as part of a broader climate adaptation strategy. Agroecology emphasizes biodiversity, soil health, and natural pest control, reducing farmers' reliance on chemical inputs and fostering more resilient farming systems. In the region of Ceará, for instance, farmers have adopted techniques such as planting drought-tolerant crops like millet and sorghum, mulching to retain soil moisture, and diversifying crops to reduce the risk of total crop failure (Marengo et al., 2022). These practices have been effective in increasing food security, improving soil quality, and ensuring a more sustainable source of income for smallholder farmers, demonstrating the resilience of agroecological approaches to climate change.

In India, the Zero Budget Natural Farming (ZBNF) model, developed by Subhash Palekar, has gained traction as a climate-resilient farming system (Government of India, 2021). ZBNF emphasizes the use of locally available resources, such as cow dung, cow urine, and natural pesticides, to minimize costs and reduce dependence on synthetic fertilizers and pesticides. Through the adoption of ZBNF, farmers in parts of Andhra Pradesh, Karnataka, and Kerala have been able to reduce their input costs, increase crop resilience, and improve soil health. The adoption of ZBNF has been particularly effective in regions prone to erratic rainfall patterns and rising temperatures, as it enhances the soil's capacity to retain moisture and withstand drought conditions. According to studies, farmers practicing ZBNF have seen significant increases in yields of staple crops like rice and vegetables, even in the face of climatic stressors (Duddigan et al., 2022). This success underscores how low-cost, sustainable farming methods can play a crucial role in adapting to climate change in the Global South.

In Egypt, a country where agriculture is heavily reliant on the Nile River, water scarcity and salinization of soil are major challenges due to climate change. To combat these issues, Egypt has invested in innovative irrigation techniques, such as drip irrigation and the reuse of treated wastewater for agricultural purposes. These technologies have helped farmers increase water-use efficiency and reduce the impact of droughts. A notable example is the introduction of solar-powered irrigation systems in Egypt's North Sinai region, where farmers have been able to access water for irrigation even during periods of low rainfall. The use of solar pumps has reduced dependency on fossil fuels, lowered irrigation costs, and improved productivity (Popoola, 2022). This initiative demonstrates the potential for renewable energy to support climate-resilient agriculture, providing farmers with a sustainable source of irrigation.

Despite the success of these adaptation projects, numerous challenges remain, particularly in terms of farmers' ability to access new technologies and markets. These challenges can impede the widespread adoption of climate-smart practices and hinder the scaling up of successful projects. Access to modern technologies remains a significant barrier for smallholder farmers in the Global South. While there are numerous climate-resilient technologies available, such as drought-resistant seed varieties, irrigation systems, and weather forecasting tools, many farmers are unable to afford or access these innovations. According to the Food and Agriculture Organization (FAO), access to inputs such as improved seeds and fertilizers is limited for many small-scale farmers, especially those in rural areas with poor infrastructure (FAO, 2020). Moreover, farmers often lack the technical knowledge to effectively implement and manage these technologies. For example, while solar-powered irrigation systems have been successful in regions like North Sinai, the high initial cost of installation and the need for maintenance skills can deter farmers from adopting them. In many cases, farmers are also unaware of the availability of such technologies due to poor dissemination of information. Another significant barrier to agricultural adaptation is the lack of access to markets. Farmers in many parts of the Global South, particularly in remote areas, face difficulty in selling their produce at fair prices due to inadequate transportation, limited market infrastructure, and the dominance of middlemen who take a large share of the profits. This market access issue is particularly acute for smallholder farmers who may lack the bargaining power to negotiate better prices for their goods. Climate change also exacerbates market instability by affecting the supply and demand of agricultural products. Fluctuating yields due to erratic weather patterns can lead to surpluses or shortages, which further destabilize prices. In Sub-Saharan Africa, for example, the agricultural sector is often characterized by low productivity, high post-harvest losses, and limited access to formal markets (Binge et al., 2023; Ariong et al., 2023; Baidhe et al., 2024; Urugo et al., 2024). These factors reduce farmers' incentives to adopt new technologies that could improve productivity if they do not have reliable access to profitable markets. Financial barriers also hinder the adoption of climate adaptation strategies. Many farmers in the Global South face difficulties in securing financing for new technologies or adaptation measures due to limited access to credit and insurance. Microfinance institutions, while growing in number, often charge high interest rates, which can discourage farmers from taking out loans for investments in new technologies. Additionally, there is often a lack of insurance products that can help farmers cope with the financial risks associated with climate-related shocks. Risk aversion is another challenge. Many farmers, especially those with limited resources, are reluctant to adopt new technologies without guarantees of success. This is particularly true when technologies require significant upfront investment or entail uncertain returns, such as new crop varieties or advanced irrigation systems. Without financial safety nets, farmers may be reluctant to experiment with climate-resilient practices.

5.5 Policy Recommendations for Agricultural Resilience in the Global South

Agriculture plays a vital role in the economies of the Global South, contributing significantly to livelihoods, food security, and poverty alleviation. However, the agricultural sector in these regions is increasingly vulnerable to climate change, market fluctuations, and socio-political instability. Therefore, promoting agricultural resilience—defined as the ability of agricultural systems to anticipate, respond to, and recover from shocks, is crucial for sustainable development. Effective national and international policies are necessary to foster agricultural resilience through key areas such as finance, infrastructure, rural development, and others.

5.5.1 *Financial Support and Access to Credit*

One of the primary barriers to enhancing agricultural resilience in the Global South is the limited access to financial resources. Smallholder farmers, who make up a substantial portion of the agricultural workforce in many developing countries, often face difficulty accessing credit, insurance, and other financial tools that are essential for managing risks associated with climate change, market volatility, and unexpected shocks. As recommendations, there should be increased public investment, support for agricultural insurance programs, and access to green finance. Governments in the Global South should prioritize agricultural financing in national budgets, ensuring that funds are allocated for infrastructure, rural development, research, and innovation. Public funding should focus on promoting smallholder farmer access to credit, ensuring that financial institutions provide affordable loans and microcredit tailored to their needs. Governments should work with international partners to develop and expand crop insurance schemes to protect farmers from climate-related risks, such as droughts and floods. These insurance programs can help stabilize farm incomes and provide a safety net in times of crisis. International financial institutions, such as the World Bank, the Green Climate Fund, and bilateral donors, can offer concessional loans and grants to support agricultural projects that focus on climate resilience, sustainable land management, and the adoption of climate-smart agricultural practices. Incentivizing private-sector investment in climate-resilient agriculture through green bonds and impact investing could further bolster financial support.

5.5.2 *Infrastructure Development*

The lack of adequate infrastructure is another significant challenge to agricultural resilience in the Global South. Poor roads, limited access to irrigation, and inadequate storage facilities result in inefficiencies in food production, distribution, and market access, often leading to food loss and decreased economic returns for farmers. As recommendations, there should be improved rural infrastructure, water management systems, and renewable energy solutions. Governments should prioritize rural infrastructure development, focusing on roads, electricity, irrigation systems, and storage facilities. This would facilitate farmers' access to markets, reduce post-harvest losses, and improve the overall productivity of agricultural systems. Public–private partnerships could be instrumental in addressing infrastructure gaps. In regions prone to water scarcity, governments should invest in efficient irrigation systems, rainwater harvesting, and sustainable water management techniques. This will help mitigate the impact of droughts and improve crop yields. Additionally, integrating traditional knowledge with modern technology could optimize water use in agriculture. Access to reliable energy sources can significantly enhance agricultural productivity. Policies should promote the use of renewable energy solutions, such as solar-powered irrigation systems or off-grid electricity for rural communities. These innovations will reduce reliance on fossil fuels and lower energy costs for farmers.

5.5.3 *Rural Development and Capacity-Building*

Rural areas in the Global South often face systemic issues such as poverty, lack of education, and limited access to essential services. These factors contribute to agricultural vulnerability by reducing the ability of farmers to adapt to changing environmental and economic conditions. Rural development policies should focus on improving livelihoods, access to services, and the capacity of farmers to manage risks and opportunities. As recommendations, there should be more investment in education and training, strengthening farmer cooperatives and networks, and social protection programs. Governments and international organizations should invest in agricultural education and capacity-building programs for farmers, especially women and youth. This includes providing training on sustainable agricultural practices, climate change adaptation strategies, and financial literacy. Empowering farmers with knowledge and skills will enable them to better manage risks, increase productivity, and adapt to climate change. Policies should encourage the formation and strengthening of farmer cooperatives and networks. These organizations can provide smallholder farmers with collective bargaining power, better access to markets, inputs, and financial resources. They can also facilitate knowledge-sharing and help farmers adopt best practices. Governments should implement social protection policies that provide safety nets for vulnerable populations in rural areas, such as food assistance, cash transfers, and public works programs. These measures can reduce the impact of shocks and provide a buffer against economic volatility.

5.5.4 *Climate-Smart Agriculture and Technology Adoption*

Climate change poses a significant challenge to agriculture in the Global South. Rising temperatures, changing rainfall patterns, and extreme weather events undermine agricultural productivity and food security. Therefore, adopting climate-smart agricultural practices is essential to build resilience. As recommendations there should be the promotion of climate-smart agricultural practices, enhancing access to agricultural technology, and research and extension services. Governments should support the widespread adoption of climate-smart agriculture (CSA) techniques that improve resilience to climate change. This includes practices such as conservation tillage, agroforestry, crop rotation, and the use of drought-resistant crop varieties. National policies should promote research and development in CSA and facilitate the dissemination of knowledge on these practices to farmers. Access to modern technology such as mobile apps for weather forecasting, precision agriculture tools, and climate-resilient seeds, can significantly improve agricultural productivity and resilience. Governments should promote policies that incentivize technology providers to offer affordable solutions to smallholder farmers, and encourage the integration of digital tools in agriculture. Governments should invest in agricultural research and extension services to develop context-specific solutions for smallholder farmers. Extension agents play a key role in disseminating new technologies, climate-resilient practices, and market information to farmers. Strengthening these services will improve farmers' capacity to adapt to new challenges.

5.5.5 *Strengthening Market Access and Trade Policies*

Agricultural resilience is not only about improving productivity but also about ensuring that farmers can access markets to sell their products at fair prices. Inadequate market access and unfair trade policies can limit farmers' income and create barriers to agricultural growth. As recommendations, there should be improvement of market infrastructure, fair trade policies, and regional trade integration. Governments should invest in market infrastructure, including local marketplaces, transportation networks, and digital platforms that connect farmers to both local and international markets. These improvements will enhance market access, reduce transaction costs, and help farmers receive fair prices for their produce. Governments and international organizations should support fair trade initiatives that ensure smallholder farmers receive equitable prices for their goods. Policy frameworks should include protections against market volatility and exploitative practices that disproportionately affect small farmers. Regional trade agreements can play a critical role in enhancing market access for farmers in the Global South. Governments should work toward creating stronger regional trade links that reduce trade barriers and promote cross-border agricultural trade.

International Cooperation and Policy Coordination
Agricultural resilience in the Global South cannot be achieved in isolation. International cooperation is essential to share knowledge, technology, and financial resources. Global and regional policy coordination can help align efforts to tackle common challenges such as climate change, food security, and market instability. As recommendations, there should be strengthening multilateral platforms and promoting south-south cooperation. International bodies such as the United Nations Food and Agriculture Organization (FAO), the World Bank, and the Green Climate Fund should continue to play a central role in fostering global dialogue and cooperation on agricultural resilience. These platforms should encourage the exchange of best practices, coordinate financial resources, and provide technical assistance to governments in the Global South. Countries in the Global South can benefit from sharing experiences, knowledge, and technologies through South-South cooperation initiatives. Governments should actively engage in regional and international forums that promote collaborative solutions to agricultural challenges.

References

AATF (2024). African Agricultural Technology Foundation. https://www.aatf-africa.org/

Adefila, A. O., Ajayi, O. O., Toromade, A. S., & Sam-Bulya, N. J. (2024). Integrating traditional knowledge with modern agricultural practices: A sociocultural framework for sustainable development. *World Journal of Biology Pharmacy and Health Sciences, 2024, 20(02), 125–135*. https://doi.org/10.30574/wjbphs.2024.20.2.0850

Ali, S. (2023). Navigating rice frontiers: A global exploration with a focus on Bangladesh's production environments. *International Journal of Agriculture and Sustainable Development, 5*(2), 90–100.

Ambikapathi, R., Schneider, K. R., Davis, B., Herrero, M., Winters, P., & Fanzo, J. C. (2022). Global food systems transitions have enabled affordable diets but had less favourable outcomes for nutrition, environmental health, inclusion and equity. *Nature Food, 3*(9), 764–779.

Antwi-Agyei, P., & Stringer, L. C. (2021). Improving the effectiveness of agricultural extension services in supporting farmers to adapt to climate change: Insights from northeastern Ghana. *Climate Risk Management, 32*, Article 100304.

Ariong, R. M., Okello, D. M., Otim, M. H., & Paparu, P. (2023). The cost of inadequate postharvest management of pulse grain: Farmer losses due to handling and storage practices in Uganda. *Agriculture & Food Security, 12*(1), 20.

Asma, J., Subrahmanyam, D., & Krishnaveni, D. (2023). The global lifeline: A staple crop sustaining twothirds of the world's population. *Agriculture Archives* 2(3): 15–18. https://doi.org/10.51470/AGRI.2023.2.3.15

Baidhe, E., Clementson, C. L., Senyah, J., & Hammed, A. (2024). Appraisal of post-harvest drying and storage operations in Africa: Perspectives on enhancing grain quality. *AgriEngineering, 6*(3), 3030–3057.

Balogun, A. L., Adebisi, N., Abubakar, I. R., Dano, U. L., & Tella, A. (2022). Digitalization for transformative urbanization, climate change adaptation, and sustainable farming in Africa: Trend, opportunities, and challenges. *Journal of Integrative Environmental Sciences, 19*(1), 17–37.

Bekuma, G. M. (2024). Climate-smart agriculture and agricultural diversification effects on productivity and resilience of smallholder farmers in Ethiopia. *International Journal of Energy and Environmental Science, 9*(6), 130–142.

Binge, B., Jalango, D., & Tesfaye, L. (2023). *Post-harvest losses management through climate smart innovations: A collaborative approach among value chain actors* (p. 53). AICCRA technical report. Accelerating Impacts of CGIAR Climate Research for Africa (AICCRA).

Duddigan, S., Collins, C. D., Hussain, Z., Osbahr, H., Shaw, L. J., Sinclair, F., Sizmur, T., Thallam, V., & Ann Winowiecki, L. (2022). Impact of zero budget natural farming on crop yields in Andhra Pradesh, SE India. *Sustainability, 14*(3), 1689.

Erezi, E., Ehi, O. E., & Ayodeji, O. T. (2023). Promoting sustainable agriculture and climate resilience in African nations. *The International Journal of Agricultural Science, 9*, 27–45.

FAO. (2020). *The state of food and agriculture 2020.* Overcoming water challenges in agriculture. https://doi.org/10.4060/cb1447en

Gebre, G. G., Amekawa, Y., & Fikadu, A. A. (2023). Farmers′ use of climate change adaptation strategies and their impacts on food security in Kenya. *Climate Risk Management, 40*, Article 100495.

Government of India (2021). *Zero budget natural farming*. https://pib.gov.in/FactsheetDetails.aspx?Id=148598&reg=3&lang=1

Hussein, A. (2024). Climate smart agriculture strategies for enhanced agricultural resilience and food security under a changing climate in Ethiopia. *Sustainable Environment, 10*(1), 2345433.

ICRAF. (2015). *Agroforestry among Kenyan farmers' tactics for climate change adaptation*. https://worldagroforestry.org/blog/2015/08/18/agroforestry-among-kenyan-farmers-tactics-for-climate-change-adaptation

IFPRI. (2024). *Southern Africa drought: Impacts on maize production*. https://www.ifpri.org/blog/southern-africa-drought-impacts-maize-production/

Izuchukwu, A., Erezi, E., & David Emeka, E. (2023). Assessing the impact of farmer-to-farmer communication networks on knowledge sharing and adoption of sustainable agricultural practices in Africa. *The International Journal of Agricultural Science, 9*, 58–76.

Jinger, D., Kumar, R., Kakade, V., Dinesh, D., Singh, G., Pande, V. C., Bhatnagar, P. R., Rao, B. K., Vishwakarma, A. K., Kumar, D., & Singhal, V. (2022). Agroforestry for controlling soil erosion and enhancing system productivity in ravine lands of Western India under climate change scenario. *Environmental Monitoring and Assessment, 194*(4), 267.

Kalele, D. N., Ogara, W. O., Oludhe, C., & Onono, J. O. (2021). Climate change impacts and relevance of smallholder farmers' response in arid and semi-arid lands in Kenya. *Scientific African, 12*, Article e00814.

Kogo, B. K., Kumar, L., & Koech, R. (2021). Climate change and variability in Kenya: A review of impacts on agriculture and food security. *Environment, Development and Sustainability, 23*(1), 23–43.

Krell, N. T. (2021). *Impacts of climate variability, decision-making and digital information on agricultural outcomes in sub-Saharan Africa*. University of California.

Lamptey, S. (2022). Agronomic practices in soil water management for sustainable crop production under rain fed agriculture of Drylands in Sub-Sahara Africa. *African Journal of Agricultural Research, 18*(1), 18–26.

Lottering, S., Mafongoya, P., & Lottering, R. (2021). Drought and its impacts on small-scale farmers in sub-Saharan Africa: A review. *South African Geographical Journal, 103*(3), 319–341.

Makate, C., Angelsen, A., Holden, S. T., & Westengen, O. T. (2022). Crops in crises: Shocks shape smallholders' diversification in rural Ethiopia. *World Development, 159*, Article 106054.

Marambanyika, V. (2022). *The role of conservation agriculture in improving smallholder farmers' crop yields in Malawi* (Master's thesis, Norwegian University of Life Sciences, Ås).

Marengo, J. A., Galdos, M. V., Challinor, A., Cunha, A. P., Marin, F. R., Vianna, M. D. S., Alvala, R. C. S., Alves, L., Moraes, O. L., & Bender, F. (2022). Drought in Northeast Brazil: A review of agricultural and policy adaptation options for food security. *Climate Resilience and Sustainability, 1*(1), Article e17.

Mbah, M., Ajaps, S., & Molthan-Hill, P. (2021). A systematic review of the deployment of indigenous knowledge systems towards climate change adaptation in developing world contexts: Implications for climate change education. *Sustainability, 13*(9), 4811.

Milhomens, A., Ávila, M. L., & de Lima Caldas, E. (2021). Agroecology and family farming: Vulnerabilities, resilience, and adaptation to climate change in the Brazilian semi-arid region. *Public Policies for Adapting Agriculture to Climate Change in Semi-Arid Northeast Brazil, 22640*, 45.

Ministry of Agriculture and Farmers Welfare (2018). *National mission for sustainable agriculture*. https://pib.gov.in/PressReleasePage.aspx?PRID=1556469

Mohammed, K., Batung, E., Kansanga, M., Nyantakyi-Frimpong, H., & Luginaah, I. (2021). Livelihood diversification strategies and resilience to climate change in semi-arid northern Ghana. *Climatic Change, 164*, 1–23.

Montesclaros, J. M. L., & Teng, P. S. (2021). Agriculture and food security in Asia. *Climate change, disaster risks, and human security: Asian experience and perspectives*, 137–168. Springer.

Mpala, T. A., & Simatele, M. D. (2024). Climate-smart agricultural practices among rural farmers in Masvingo district of Zimbabwe: Perspectives on the mitigation strategies to drought and water scarcity for improved crop production. *Frontiers in Sustainable Food Systems, 7*, 1298908.

Msongaleli, B. (2024). The contribution of climate-smart agriculture to reducing climate-related risks to rain-fed maize production: Insights from Tanzania's semi-arid and sub-humid regions. *Ghana Journal of Geography, 16*(1), 28–58.

Nayak, S., Das, K., Bhowmick, M. K., Hossain, S. M., Singh, S., Singh, U. M., & Singh, V. K. (2022). Flood stress prevalence and assessment of farmers' preparedness with resilient rice varieties: A review in context of Eastern India. *Response of Field Crops to Abiotic Stress*, 277–296.

Neupane, N., Chaudhary, P., Rijal, Y., Ghimire, B., & Bhandari, R. (2022). The role of renewable energy in achieving water, energy, and food security under climate change constraints in South Asia. *Frontiers in Sustainable Food Systems, 6*, 1016093.

Ngcamu, B. S. (2023). Climate change effects on vulnerable populations in the Global South: A systematic review. *Natural Hazards, 118*(2), 977–991.

Odhiambo, K. O., Iro Ong'or, B. T., & Kanda, E. K. (2021). Optimization of rainwater harvesting system design for smallholder irrigation farmers in Kenya: A review. *AQUA—Water Infrastructure, Ecosystems and Society*, *70*(4), 483–492.

Panda, D. K., Tiwari, V. M., & Rodell, M. (2022). Groundwater variability across India, under contrasting human and natural conditions. *Earth's Future*, *10*(4), e2021EF002513.

Pandey, S. C., Modi, P., Pereira, V., & Fosso Wamba, S. (2024). Empowering small farmers for sustainable agriculture: A human resource approach to SDG-driven training and innovation. *International Journal of Manpower*. https://doi.org/10.1108/IJM-11-2023-0655

Popoola, O. D. (2022). *A nexus approach to solar pumping irrigation systems in North Africa: opportunities and challenges: a case study of Egypt*. http://essay.utwente.nl/92614/1/Popoola_MA_BMS.pdf

Porkka, M., Wang-Erlandsson, L., Destouni, G., Ekman, A. M., Rockström, J., & Gordon, L. J. (2021). Is wetter better? Exploring agriculturally-relevant rainfall characteristics over four decades in the Sahel. *Environmental Research Letters, 16*(3), Article 035002.

Raj, S., Roodbar, S., Brinkley, C., & Wolfe, D. W. (2022). Food security and climate change: Differences in impacts and adaptation strategies for rural communities in the global south and north. *Frontiers in Sustainable Food Systems, 5*, Article 691191.

Saley, I. A., & Salack, S. (2023). Present and future of heavy rain events in the Sahel and West Africa. *Atmosphere, 14*(6), 965.

Samuel, J., Rao, C. A. R., Raju, B. M. K., Reddy, A. A., Pushpanjali, R., & A. G. K., Kumar, R. N., Osman, M., Singh, V. K., & Prasad, J. V. N. S. (2021). Assessing the impact of climate resilient technologies in minimizing drought impacts on farm incomes in drylands. *Sustainability, 14*(1), 382.

Sanogo, M., Yonaba, R., Lawane, A., Zorom, M., Tassembédo, F., Ali Sahad, H., & Bazié, I. (2024). Do Runoff water harvesting ponds affect farmers cropping choices? Insights from smallholders in the West African Sahel. *Sustainability, 16*(18), 8000.

Siatwiinda, S. M., Supit, I., van Hove, B., Yerokun, O., Ros, G. H., & de Vries, W. (2021). Climate change impacts on rainfed maize yields in Zambia under conventional and optimized crop management. *Climatic Change, 167*(3), 39.

Stratton, A. E., Wittman, H., & Blesh, J. (2021). Diversification supports farm income and improved working conditions during agroecological transitions in southern Brazil. *Agronomy for Sustainable Development, 41*(3), 35.

Sultana, N., & Luetz, J. M. (2022). Adopting the local knowledge of coastal communities for climate change adaptation: A case study from Bangladesh. *Frontiers in Climate, 4*, Article 823296.

Tembo, A., Muyabe, O., Musenge, D. C., Mhango, J., & Nkomanga, G. C. (2025). Impact of sustainable agricultural practices on farm productivity, yield, and climate resilience among smallholder farmers in Zambia. *Journal of Agriculture, Aquaculture, and Animal Science, 2*(1), 30–38.

Turner, M. D., & Ayantunde, A. (2022). Household diversification and market dependence: Understanding vulnerability in rural West Africa. *The Journal of Peasant Studies, 49*(7), 1497–1523.

UNDRR (2024). *Horn of Africa floods and drought, 2020–2023—Forensic analysis*. https://www.undrr.org/resource/horn-africa-floods-and-drought-2020-2023-forensic-analysis

Urugo, M. M., Yohannis, E., Teka, T. A., Gemede, H. F., Tola, Y. B., Forsido, S. F., Tessema, A., Suraj, M., & Abdu, J. (2024). Addressing post-harvest losses through agro-processing for sustainable development in Ethiopia. *Journal of Agriculture and Food Research*, 101316.

Verma, A., Kumar, P., & Saresh, N. V. (2021). Secondary metabolites: Harvesting short term benefits from arid zone agroforestry systems in India. *Agroforestry Systems, 95*, 515–532.

Villalba, R., Joshi, G., Daum, T., & Venus, T. E. (2024). Financing climate-smart agriculture: A case study from the Indo-Gangetic plains. *Mitigation and Adaptation Strategies for Global Change, 29*(5), 33.

Vrielink, M. (2023). *Silent ideologies and power relations: a critical discourse analysis of ICT4D initiatives within the Kenyan agricultural sector* (Master's thesis).

Wanjira, J. K. (2021). *Adoption of climate-smart maize varieties and its impact on household income among small-scale farmers in Embu County, Kenya* (Doctoral dissertation, UON).

Waqas, M., Naseem, A., Humphries, U. W., Hlaing, P. T., Shoaib, M., & Hashim, S. (2024). A comprehensive review of the impacts of climate change on agriculture in Thailand. *Farming System*, 100114.

Wens, M. L., Van Loon, A. F., Veldkamp, T. I., & Aerts, J. C. (2022). Education, financial aid, and awareness can reduce smallholder farmers' vulnerability to drought under climate change. *Natural Hazards and Earth System Sciences, 22*(4), 1201–1232.

World Bank (2020). *Revised estimates of the impact of climate change on extreme poverty by 2030* (Policy Research Working Paper 9417). https://documents1.worldbank.org/curated/en/706751601388457990/pdf/Revised-Estimates-of-the-Impact-of-Climate-Change-on-Extreme-Poverty-by-2030.pdf

World Health Organization. (2021). *The state of food security and nutrition in the world 2021: Transforming food systems for food security, improved nutrition and affordable healthy diets for all* (p. 2021). Food & Agriculture Organization of the United Nations.

Yuan, X., Li, S., Chen, J., Yu, H., Yang, T., Wang, C., Huang, S., Chen, S., & Ao, X. (2024). Impacts of global climate change on agricultural production: A comprehensive review. *Agronomy, 14*(7), 1360.

Yuen, K. W., Switzer, A. D., Teng, P. P., & Lee, J. S. H. (2022). Assessing the impacts of tropical cyclones on rice production in Bangladesh, Myanmar, Philippines, and Vietnam. *Natural Hazards and Earth System Sciences Discussions, 2022*, 1–28.

Zakari, S., Ibro, G., Moussa, B., & Abdoulaye, T. (2022). Adaptation strategies to climate change and impacts on household income and food security: Evidence from Sahelian region of Niger. *Sustainability, 14*(5), 2847.

CHAPTER 6

Climate Resilience Policies in the Fishery, Agriculture, Pastoralist, and Forest Sectors in the Developing World: State of the Art and Perspectives

Abstract This chapter provides a comprehensive analysis of climate resilience policies across the fishery, agriculture, pastoralist, and forest sectors in the Global South. It explores the current landscape of climate resilience strategies, focusing on the integration of local knowledge and community input to ensure relevance and effectiveness. The importance of ecosystem-based adaptation (EbA) and climate-smart agriculture (CSA) is emphasized as emerging solutions that align with sustainable development goals and promote green economies. The chapter also highlights the challenges faced in policy implementation, including inadequate funding, political inertia, and the lack of inter-sectoral coordination. These barriers often undermine the effectiveness of climate resilience measures. It equally outlines future policy directions, advocating for inclusive, flexible, and comprehensive frameworks that not only recognize sectoral challenges but also integrate local realities and traditional knowledge. Through a focus on cross-sectoral cooperation and the promotion of innovative approaches, the chapter presents strategies for strengthening climate resilience in the Global South.

Keywords Climate resilience · Global South · Local knowledge · Ecosystem-based adaptation · Climate-smart agriculture · Policy integration · Green economies

N. P. Awazi, *Building Resilience*, Palgrave Studies in Climate Resilient Societies, https://doi.org/10.1007/978-3-031-93791-0_6

6.1 Current Climate Resilience Policy Landscape in the Fishery, Agriculture, Pastoralist, and Forest Sectors in the Global South

Climate change poses significant challenges to the Global South, where communities dependent on natural resources such as fisheries, agriculture, pastoralism, and forests are particularly vulnerable. The interconnectedness between these sectors and the livelihoods of millions underscores the urgency of adapting to the changing climate. Resilience policies aim to enhance the capacity of vulnerable communities and ecosystems to cope with and recover from climate-induced shocks. Given the importance of these sectors to both national economies and individual livelihoods, understanding climate resilience strategies is critical for long-term sustainable development in the Global South (Table 6.1).

The fisheries sector in the Global South is highly vulnerable to the impacts of climate change, particularly due to changes in water temperature, ocean acidification, altered fish migration patterns, and extreme weather events (Cinner et al., 2022; Selvaraj et al., 2022). These changes have direct consequences for food security, income generation, and cultural practices in coastal and island communities. In response to these challenges, many countries in the Global South have adopted various climate resilience policies. The United Nations Framework Convention on Climate Change (UNFCCC) emphasizes the importance of sustainable fisheries management in adaptation strategies. Several countries have implemented policies to promote the sustainable management of marine resources. For example, in Sub-Saharan Africa, the "African Adaptation Initiative" encourages fisheries management practices that incorporate climate risks into local decision-making (AAI, 2024; UNDP, 2025). Additionally, climate-smart fisheries management, which focuses on enhancing adaptive capacity and reducing vulnerability, has gained attention. Key policy instruments include establishing marine protected areas (MPAs), improving governance structures, and promoting sustainable fishing techniques (White et al., 2022). In coastal areas, policies focus on providing alternative livelihoods to communities dependent on fisheries. The Intergovernmental Panel on Climate Change (IPCC) underscores the need for integrated policies that address both ecological and socio-economic aspects of climate resilience in the fisheries sector (IPCC, 2022). In countries like India, the government has initiated policies like the National Fisheries Policy, which incorporates climate resilience strategies such as

Table 6.1 Climate resilience policy landscape in the fisheries, agriculture, pastoralist, and forest sectors in the Global South

Sector	*Climate resilience policies*	*Key examples and challenges*	*Key references*
Fisheries	Policies focus on sustainable fisheries management, climate-smart fishing techniques, and alternative livelihoods. Key strategies include marine protected areas (MPAs), better governance, and adaptive capacity enhancement	– **Sub-Saharan Africa**: The African Adaptation Initiative promotes fisheries management that incorporates climate risks – **India**: National Fisheries Policy includes climate resilience strategies for sustainable aquaculture and disaster preparedness.-Fragmented governance hampers effective adaptation	Selvaraj et al. (2022), Cinner et al. (2022), AAI, (2024), UNDP (2025), White et al. (2022), IPCC (2022), and Katznelson et al. (2025)
Agriculture	Climate-Smart Agriculture (CSA) aims to increase productivity, adapt to climate impacts, and reduce greenhouse gas emissions. Policies promote crop diversification, irrigation improvements, and agroforestry	– **Kenya and Ghana**: Policies emphasize crop diversification, irrigation technologies, and early warning systems – **Adaptation Fund** supports agricultural projects focusing on smallholder resilience - Barriers include limited finance, technology gaps, and weak institutional support	Lipper et al. (2014), Frimpong et al. (2023), Aduhene-Chinbuah et al. (2024), Nguyen et al. (2021), and Mungai et al. (2022)

(continued)

Table 6.1 (continued)

Sector	*Climate resilience policies*	*Key examples and challenges*	*Key references*
Pastoralism	Policies focus on restoring degraded lands, improving water access, enhancing livestock mobility, and incorporating local knowledge in climate predictions	– **Ethiopia**: PCDP III improves drought early warning systems and water harvesting – **Africa**: Pastoralist National Adaptation Strategy includes land restoration and enhancing mobility.- Conflicts over land and water resources and the legal recognition of pastoralismremain significant challenges	ILRI (2025), IFAD (2019), UNEP (2025a, 2025b), and Nori and Scoones (2023)
Forests	REDD+ initiatives aim to reduce deforestation and enhance carbon storage. Policies also promote agroforestry, sustainable forest management, and community-based forest management	– **Brazil and Indonesia**: REDD + programs focus on emissions reduction and poverty alleviation – **Nepal and Costa Rica**: Community-based forest management supports conservation and provides income through ecotourism and timber.- Enforcement of policies remains a key challenge	Gakou Kakeu (2022), Sacherer et al. (2022), Koh et al. (2024), World Bank (2025), and Chichinye et al. (2024)

sustainable aquaculture practices and disaster preparedness (National Fisheries Policy of India, 2020). However, the implementation of these policies has been challenging due to limited resources, lack of capacity at local levels, and insufficient integration of scientific research into decision-making. A key challenge remains the fragmentation of fisheries management at different levels of governance, which hampers effective adaptation to climate change (Katznelson et al., 2025). The overfishing of marine ecosystems coupled with climate stressors exacerbates the vulnerability of small-scale fisheries, making climate adaptation policies even more critical.

Agriculture is the backbone of many economies in the Global South, providing livelihoods for a majority of the population. However, climate change is intensifying challenges such as droughts, floods, and unpredictable weather patterns. This has significant implications for crop yields, food security, and rural development. In response to these challenges, countries in the Global South are implementing climate-resilient agricultural practices through policy frameworks such as the "Climate-Smart Agriculture(CSA) approach promoted by the Food and Agriculture Organization (FAO) of the United Nations. CSA seeks to enhance food security and agricultural sustainability in the context of climate change by focusing on three pillars: increasing productivity, adapting to climate impacts, and reducing greenhouse gas emissions (Lipper et al., 2014). Policies in countries such as Kenya and Ghana emphasize crop diversification, improved irrigation technologies, and early warning systems to combat the impact of climate change on agriculture (Aduhene-Chinbuah et al., 2024; Frimpong et al., 2023). These policies are closely tied to the broader goals of sustainable development and poverty reduction, which are central to climate resilience. In addition, the "Adaptation Fund," a financial mechanism under the Kyoto Protocol, has supported agricultural projects in the Global South, particularly in regions like Latin America, South Asia, and Sub-Saharan Africa. These projects focus on improving the resilience of smallholder farmers through climate-resilient crop varieties, better soil management practices, and agroforestry. Studies by Nguyen et al. (2021) and Mungai et al. (2022) note that the main barriers to effective climate adaptation in agriculture include limited access to finance, technology gaps, and the absence of strong institutional support. Furthermore, climate variability, rather than climate change itself, often exacerbates the vulnerability of agricultural systems. In many cases, farmers lack access to accurate climate information and early warning systems, which hinders their ability to adapt.

Pastoralism, a way of life for millions of people in the Global South, is heavily impacted by climate change. Changes in rainfall patterns, increasing temperatures, and the degradation of grazing lands directly affect the livelihoods of pastoralist communities. In Africa, Asia, and parts of Latin America, pastoralists are struggling with more frequent and prolonged droughts and floods, which undermine their ability to maintain livestock. In response, policies have emerged to support the resilience of pastoralist systems. One prominent framework is the "Pastoralist National Adaptation Strategy," adopted in several African countries. This strategy includes a range of policy measures such as the restoration of degraded land, improving access to water, and enhancing the mobility of livestock. The International Livestock Research Institute (ILRI) has advocated for policies that integrate climate science and local knowledge to better predict weather patterns and manage grazing systems (ILRI, 2025). In Ethiopia, the "Pastoral Community Development Project III" (PCDP III) has successfully implemented climate resilience policies through the improvement of drought early warning systems, water harvesting techniques, and community-based resource management (IFAD, 2019). These policies aim to empower pastoralists to make informed decisions regarding herd management, improve livestock productivity, and reduce vulnerability to climate shocks. One of the critical challenges facing the pastoralist sector is the increasing conflict over land and water resources. As water scarcity increases due to climate change, pastoralists face competition from agriculturalists and urban areas. Climate-induced migration also exacerbates tensions between communities. According to the United Nations Environment Programme (UNEP), conflict prevention and conflict-sensitive policies are necessary to promote peace and resilience in pastoralist regions (UNEP, 2025a, 2025b). Additionally, policy gaps exist in the formal recognition of pastoralism as a sustainable land-use practice. In many countries, policies tend to prioritize sedentary agriculture over pastoralism, leading to land use and land tenure issues for pastoralists (Nori & Scoones, 2023). Strengthening the legal framework around land rights and ensuring inclusive participation in policy development are key to fostering climate resilience in pastoralist communities.

Forests are critical for climate resilience due to their role in carbon sequestration, biodiversity conservation, and regulating local weather patterns. In the Global South, forests support millions of people directly through timber, non-timber products, and ecosystem services, and indirectly through the role they play in mitigating climate change. Climate resilience policies in the forest sector focus on reducing deforestation, promoting sustainable forest management, and enhancing carbon storage. One prominent policy initiative is REDD+ (Reducing Emissions from Deforestation and Forest Degradation), which encourages forest conservation by providing financial incentives to countries that reduce emissions from deforestation. Countries like Brazil, Indonesia, Cameroon, Democratic Republic of Congo and Republic of Congo have implemented REDD+ programs, which combine climate action with poverty reduction and biodiversity conservation goals (Gakou Kakeu, 2022; Koh et al., 2024; Sacherer et al., 2022). Another key policy response is the promotion of agroforestry, which integrates trees into agricultural landscapes to improve soil fertility, water retention, and biodiversity. The World Bank's "Forest and Landscape Restoration" program has supported countries in the Global South to implement restoration initiatives that enhance forest cover and strengthen ecosystem resilience to climate change (World Bank, 2025). In countries like Nepal and Costa Rica, community-based forest management has emerged as an important climate resilience strategy. Local communities manage forests, which not only helps preserve biodiversity but also provides income through ecotourism and sustainable timber production. However, challenges persist in ensuring equitable distribution of benefits, particularly for marginalized communities. Despite these efforts, forest loss continues at alarming rates due to illegal logging, agricultural expansion, and infrastructure development. According to Chichinye et al. (2024), the slow pace of policy implementation and the lack of enforcement mechanisms are significant barriers to progress in forest conservation in many parts of the Global South.

6.2 Policy Integration with Local Knowledge for Climate Resilience in the Fishery, Agriculture, Pastoralist, and Forest Sectors in the Global South

Climate change poses significant threats to the livelihoods and ecosystems of communities in the Global South, particularly in sectors such as fisheries, agriculture, pastoralism, and forestry. Given the increasing intensity of climate impacts in these regions, building resilience to climate change is of paramount importance. A promising approach for addressing this challenge is integrating local knowledge, particularly traditional ecological knowledge (TEK), into climate resilience strategies.

Local knowledge, often developed over centuries by indigenous and local communities, is a key resource for understanding environmental processes and risks. It encompasses a deep understanding of ecosystems, weather patterns, agricultural practices, animal husbandry, and resource management, all shaped by lived experiences and passed down through generations. In the context of climate resilience, local knowledge plays a critical role by providing insights into how communities have historically adapted to climatic variations and extreme events. Integrating this knowledge into climate change adaptation strategies can enhance the effectiveness, relevance, and sustainability of these efforts. The integration of local knowledge into policy is particularly important in the Global South, where many communities have already experienced significant climatic fluctuations and are highly vulnerable to future impacts (Iwama et al., 2021). As a result, these communities possess a wealth of experiential knowledge about how to manage resources and cope with changing environmental conditions. Recognizing and valuing this knowledge is essential to developing policies that are not only scientifically sound but also socially and culturally appropriate.

In the fisheries sector, local knowledge has long been used to manage fish populations and sustain fishing practices. Many coastal and inland communities in the Global South rely on fish for food security, income, and cultural practices. Local fishers possess intricate knowledge of fish species, breeding cycles, migration patterns, and seasonal variations in water bodies. This knowledge is crucial for understanding how fish populations may be impacted by climate change, including the warming of oceans, changes in rainfall patterns, and increased frequency of extreme weather events. For example, in the Indian Ocean, small-scale fishers have developed intricate systems of observing oceanographic conditions, such as the color of the sea and the presence of certain types of seaweed, to predict the movement of fish. However, climate change is altering these patterns, and many fishers report changes in the timing and location of fish migrations. Integrating this local knowledge with scientific climate data can create more effective management strategies that balance sustainability with community needs. A successful example of such integration can be found in the Pacific Islands, where communities have combined traditional fishing knowledge with modern science to manage marine resources. In Fiji, the concept of "locally managed marine areas" (LMMAs) has been developed, where traditional resource management practices are integrated with scientific research to create more effective conservation strategies (Fache & Breckwoldt, 2018; Singh et al., 2021). Such collaborations enable the design of adaptive strategies that account for both local knowledge and scientific insights, ultimately leading to enhanced resilience in the face of climate change.

Agriculture is one of the sectors most affected by climate change, particularly in the Global South, where smallholder farmers are highly dependent on local weather patterns for planting and harvesting. Traditional farming practices are often finely tuned to local conditions and include knowledge of soil types, crop varieties, and water management techniques that have been passed down through generations. This localized knowledge is especially valuable in the context of climate variability and change, as it enables farmers to respond to shifts in growing seasons, droughts, and floods. For instance, in parts of East Africa, farmers have developed diverse cropping systems that enable them to adapt to unpredictable rainfall patterns. They utilize indigenous crop varieties that are more resistant to drought and pests, such as millet and sorghum. These crops are often more resilient to climate variability than commercial crops like maize. In Kenya, the integration of indigenous crop varieties and agricultural practices into climate resilience strategies has proven effective in sustaining food production despite changing climatic conditions (Mulungu & Kangogo, 2022). Moreover, farmers' observations of weather patterns, such as the behavior of local birds, cloud formations, and plant blooming cycles, provide valuable predictive information that can complement meteorological data. Policies that integrate these observations with formal forecasting systems can help farmers make better decisions about when to plant, irrigate, and harvest, enhancing food security and reducing vulnerability to climate shocks.

Pastoralism is another sector that relies heavily on local knowledge for resilience. In many parts of the Global South, pastoralists manage livestock in arid and semi-arid regions, where climate change has exacerbated the challenges posed by drought, shifting grazing patterns, and reduced access to water sources. Traditional knowledge systems in pastoral communities often involve complex systems of resource management, such as transhumance (seasonal migration) and the use of drought-resistant animal breeds. In East Africa, pastoralists have long adapted to variations in rainfall and the availability of grazing land by moving their herds across vast areas in response to changing environmental conditions. This knowledge is critical to their ability to maintain healthy livestock and minimize the impacts of droughts and other climatic stresses. However, as climate change leads to more erratic rainfall patterns and prolonged dry periods, these traditional practices are being threatened. Integrating local knowledge into policy for pastoralist communities involves recognizing the importance of mobility and flexible grazing strategies, as well as maintaining access to key resources like water and pasturelands (Nori, 2022). In Kenya and Ethiopia, policies that support mobile pastoralism have been shown to be more effective in managing the impacts of climate change than those that encourage sedentary lifestyles or fixed land-use practices (Campbell, 2021; Lelenguyah, 2023).

Forestry, especially in the context of community-based forest management, can benefit greatly from the integration of local knowledge into climate resilience strategies. Indigenous and local communities have long managed forests using traditional ecological practices, such as selective logging, controlled burning, and the protection of sacred groves. These practices help to maintain biodiversity, protect soil health, and sustain ecosystem services, which are all critical in the context of climate change. In the Amazon rainforest, indigenous communities have developed agroforestry systems that combine tree planting with agricultural cultivation, creating resilient ecosystems that are better able to withstand environmental changes. These systems are highly adaptive to varying rainfall patterns and temperatures, and they also promote the conservation of forest biodiversity. Recent initiatives to integrate indigenous forest management practices into national climate strategies have shown promising results in enhancing forest resilience and mitigating climate change impacts (Baskent, 2024; Muthee et al., 2022).

Despite the significant potential benefits of integrating local knowledge into climate resilience strategies, several challenges remain. One of the main obstacles is the marginalization of indigenous and local communities in decision-making processes. National policies, often driven by scientific and top-down approaches, can overlook or dismiss the value of local knowledge. Furthermore, issues of land tenure, access to resources, and political representation can limit the ability of communities to actively participate in climate adaptation planning. However, there are increasing efforts to overcome these barriers. In many countries, governments and international organizations are recognizing the importance of participatory approaches to climate resilience. The United Nations Framework Convention on Climate Change (UNFCCC) and other global institutions have acknowledged the need to include indigenous knowledge systems in climate action, as evidenced by initiatives such as the Local Coproduction of Climate Knowledge (LCCK) and the Indigenous Peoples' Major Group.

6.3 Ecosystem-Based Adaptation and Climate-Smart Agriculture for Climate Resilience in the Global South

Climate change poses a significant challenge to the global south, particularly due to its heightened vulnerability and limited adaptive capacity. The negative impacts of climate change, including rising temperatures, extreme weather events, changing precipitation patterns, and sea-level rise, disproportionately affect developing countries. These regions often rely heavily on agriculture, forestry, and fisheries, sectors that are directly linked to environmental changes. As a result, innovative strategies for adaptation to these climate-induced challenges are crucial. Two such strategies, i.e., Ecosystem-Based Adaptation (EbA) and Climate-Smart Agriculture (CSA), offer sustainable and integrated solutions. These approaches are central to building climate resilience in the global south and fostering green economies.

Ecosystem-Based Adaptation refers to the use of biodiversity and ecosystem services to help people adapt to the adverse effects of climate change. According to the International Union for Conservation of Nature (IUCN), EbA integrates the conservation and sustainable use of ecosystems into an overall adaptation strategy (IUCN, 2022). In the global south, where ecosystems provide critical resources such as water, food, and shelter, maintaining and restoring ecosystems can offer both climate mitigation and adaptation benefits. EbA strategies are particularly pertinent for rural communities in the global south, which often depend on natural resources for their livelihoods. For instance, the restoration of mangroves along coastal areas can help protect against storm surges and rising sea levels, thus reducing vulnerability to extreme weather events. In East Africa, the restoration of degraded land and ecosystems, such as grasslands and forests, has been shown to increase water retention, improve soil fertility, and reduce the impact of droughts (Eschen et al., 2023). Similarly, the reforestation of watershed areas in countries like Honduras and Nepal has led to more reliable water supplies and less flooding (Laudari et al., 2022; Ojhaa et al., 2021). These kinds of ecosystem-based solutions have the potential to increase resilience to climate change while also providing co-benefits such as poverty alleviation, food security, and biodiversity conservation. A prominent example of EbA implementation is the "Green Belt Movement" in Kenya, which has focused on reforesting degraded lands while simultaneously addressing climate change and improving local communities' livelihoods (Purohit et al., 2024). The movement has empowered local populations, particularly women, by integrating environmental restoration with community development, showcasing a successful synergy between ecological conservation and climate resilience.

Climate-Smart Agriculture (CSA) is an approach that aims to increase agricultural productivity, enhance resilience to climate change, and reduce greenhouse gas emissions. According to the Food and Agriculture Organization (FAO), CSA involves practices, policies, and investments that help farmers increase productivity in the face of climate change, while also reducing emissions and enhancing the resilience of agricultural systems (FAO, 2025). For the global south, where agriculture constitutes a large part of the economy, CSA can significantly contribute to food security and poverty reduction while addressing the challenges posed by climate change. CSA encompasses a wide range of practices, from the adoption of drought-resistant crop varieties to agroforestry, improved water management, and the diversification of crops to increase resilience to climate shocks (Hussein, 2024). A key aspect of CSA is the integration of scientific knowledge with indigenous practices. For example, in countries like India, the use of traditional knowledge, such as crop rotation and mixed farming systems, is combined with modern techniques like precision irrigation to optimize yields while minimizing the environmental footprint of agriculture (Kakraliya et al., 2022). In Sub-Saharan Africa, CSA has been successfully implemented through initiatives such as the "Climate-Smart Villages" (CSV) project. These projects use participatory approaches to help farmers identify and implement practices that increase agricultural productivity while reducing vulnerability to climate change. In Ethiopia, for example, farmers have adopted CSA practices such as improved soil management techniques and water conservation practices, which have led to increased yields even in the face of climate variability (Zerssa et al., 2021). One notable CSA initiative is the use of "Agroecology" in Brazil, where the combination of organic farming and agroforestry systems has promoted sustainable farming while reducing deforestation in the Amazon. Agroecology in the global south is seen as a way to combat both environmental degradation and climate change while simultaneously improving food security and smallholder livelihoods (Lucantoni & Domarle, 2023). These practices show how CSA can be tailored to local contexts and contribute to the long-term resilience of farming systems.

The concept of a green economy is increasingly being championed as a path toward sustainable development in the global south. A green economy is one that promotes sustainable economic growth while ensuring the protection of natural resources and social equity. According to the United Nations Environment Programme (UNEP), a green economy is low-carbon, resource-efficient, and socially inclusive (UNEP, 2025a, 2025b). In the context of the global south, transitioning to a green economy is critical for addressing both the impacts of climate change and the region's development challenges, such as poverty, inequality, and food insecurity. Ecosystem-Based Adaptation and Climate-Smart Agriculture are integral to the green economy model because they contribute to the sustainable use of natural resources while building climate resilience. For example, the adoption of agroforestry and sustainable land management practices can sequester carbon, reduce emissions, and enhance ecosystem services. This approach aligns with the green economy principles by fostering low-carbon and climate-resilient development pathways. In the global south, the transition to a green economy also involves creating green jobs in sectors such as renewable energy, ecotourism, and sustainable agriculture, which can help lift communities out of poverty while simultaneously mitigating the effects of climate change (Murphy, 2024).

While EbA and CSA have demonstrated significant potential for enhancing climate resilience in the global south, several challenges hinder their widespread adoption. One of the primary obstacles is the lack of financial resources to scale up these approaches. Many governments in the global south face budget constraints and rely heavily on international aid for climate adaptation. As a result, there is a need for increased investment in both public and private sectors to support sustainable development and climate resilience efforts. Additionally, there are barriers related to knowledge gaps and access to technology. Many farmers and local communities in the global south lack access to climate information, training, and modern tools necessary to implement CSA and EbA practices effectively. Overcoming these barriers requires greater investment in education, extension services, and the development of climate-smart technologies that are accessible and affordable. Another significant challenge is the governance and policy environment. Fragmented policies, weak institutions, and limited coordination between sectors and stakeholders can hinder the successful implementation of EbA and CSA strategies. However, opportunities for overcoming these challenges exist.

For instance, through international partnerships and climate financing mechanisms such as the Green Climate Fund, financial support for EbA and CSA can be secured. Additionally, strengthening local institutions and involving communities in the design and implementation of adaptation strategies can enhance ownership and long-term sustainability.

6.4 Challenges in Climate Resilience Policy Implementation in the Fishery, Agriculture, Pastoralist, and Forest Sectors in the Global South

Climate change poses significant challenges to various sectors, particularly those in the Global South, which are more vulnerable to its adverse impacts. The fisheries, agriculture, pastoralist, and forest sectors are some of the most susceptible to climate change, as they directly rely on climate patterns and environmental conditions. In response to these challenges, many governments and international organizations have developed climate resilience policies aimed at minimizing the damage caused by climate change, adapting to new conditions, and fostering long-term sustainability. However, implementing these policies effectively has proven to be a daunting task due to several barriers. Common challenges include insufficient funding, lack of political will, inadequate inter-sectoral coordination, and weak institutional capacities, among others (Table 6.2).

One of the most prominent barriers to effective climate resilience policy implementation is insufficient funding. Climate resilience measures often require significant financial investment, both for research and development and for the implementation of practical solutions on the ground. However, countries in the Global South often face fiscal constraints, limiting their ability to mobilize resources for climate adaptation. In the fisheries sector, for example, adapting to climate change may require investment in sustainable fishing practices, improved monitoring and regulation systems, and the development of alternative livelihoods for affected communities (Bryndum-Buchholz et al., 2021). Similarly, agriculture, which is highly sensitive to changing precipitation patterns and extreme weather events, requires significant investment in drought-resistant crops, soil conservation techniques, and irrigation infrastructure (Marengo et al., 2022). For pastoralists, the impacts of climate change necessitate investments in water access, livestock management,

Table 6.2 Challenges in implementing climate resilience policies across the fisheries, agriculture, pastoralist, and forest sectors in the Global South

Sector	*Challenges in climate resilience policy implementation*	*Key examples and barriers*	*Key references*
Fisheries	1. **Insufficient funding**: Limited financial resources for sustainable fishing practices and community support 2. **Weak institutional capacity**: Lack of data and monitoring systems	– **Fisheries**: Insufficient investment for monitoring fish stocks, improving regulations, and developing alternative livelihoods – **Data gaps**: Lack of data on marine ecosystems to guide adaptive policies	Bryndum-Buchholz et al. (2021) and Bolin et al. (2021)
Agriculture	1. **Insufficient funding**: Limited investment in drought-resistant crops, irrigation systems, and soil conservation 2. **Weak institutional capacity**: Shortage of extension services and training	– **Agriculture**: Lack of financial resources for adaptive infrastructure – **Data gaps**: Limited access to accurate climate data for farmers	Marengo et al. (2022), Antwi-Agyei and Stringer (2021) Osumba et al. (2021)
Pastoralism	1. **Lack of political will**: Resistance from political elites and competing land and water interests 2. **Weak institutional capacity**: Poor land-use policies and access to relevant climate information 3. **Inter-sectoral coordination**: Lack of cross-sector cooperation	– **Pastoralism**: Difficulty coordinating across sectors (e.g., livestock, water, land tenure) – **Water access**: Lack of policies for improving water access and mobility – **Cross-border issues**: Coordination challenges across borders	Rasul and Neupane (2021), Postigo (2021), and Senda et al. (2022)

(continued)

Table 6.2 (continued)

Sector	*Challenges in climate resilience policy implementation*	*Key examples and barriers*	*Key references*
Forests	1. **Insufficient funding**: Limited financial resources for forest management, reforestation, and enforcement 2. **Weak political will**: Land use and logging pressures hinder forest protection 3. **Weak enforcement**: Lack of capacity to monitor illegal activities	– **Forest sector**: Underfunded initiatives for reforestation and deforestation reduction – **Political pressures**: Short-term economic interests undermining long-term forest policies – **Enforcement gaps**: Weak enforcement of forest laws	Ahmadzai et al. (2022), Wando (2023), and Hifume et al. (2024)
Cross-Sectoral Challenges	1. **Inter-sectoral coordination**: Difficulty in coordinating across different ministries (e.g., agriculture, water, forestry) 2. **Data gaps**: Lack of reliable, accurate climate data for informed decision-making	– **Cooperation failures**: Ministries operating in silos, leading to inefficiencies – **Data shortages**: Poor data on local climate patterns and climate impacts	Khalid (2023), Nemakonde and Van Niekerk (2023) Aryal et al. (2021), and Smidt and Jokonya (2022)

and mobility systems to adjust to changing grazing patterns (Postigo, 2021). Forest sector adaptation, likewise, requires funding for forest management, reforestation efforts, and measures to reduce deforestation (Ahmadzai et al., 2022). Despite the need for these investments, many countries in the Global South face challenges in securing adequate funding from both domestic and international sources. Even when funds are available, they are often insufficient or not effectively targeted at the most vulnerable communities. This lack of funding limits the scale of interventions that can be implemented and undermines the effectiveness of climate resilience policies.

Another significant barrier to climate resilience policy implementation is the lack of political will. While many governments in the Global South acknowledge the risks posed by climate change and have committed to climate resilience policies, political inertia often prevents these policies from being effectively enacted or implemented. Political will is particularly crucial in sectors like agriculture, fisheries, and forestry, where policy changes may disrupt entrenched practices and interests. In the agriculture sector, for instance, governments may face resistance from powerful agricultural lobbies or political elites who are reluctant to change existing practices or invest in long-term adaptation measures (Resnick, 2024). Similarly, in the forest sector, political pressures related to land use, logging interests, and the need for immediate economic returns can undermine efforts to implement policies that protect forests and enhance resilience (Wando, 2023). Political leaders may also be swayed by short-term electoral considerations, which often prioritize immediate economic growth over long-term sustainability and resilience. In addition, weak governance structures and a lack of accountability can contribute to political apathy or lack of action. Even when policies are formulated, poor enforcement mechanisms or corruption can prevent their effective implementation. Without strong political will to prioritize climate resilience, policies are likely to remain ineffective, and vulnerable communities will continue to face the brunt of climate impacts.

Climate resilience policies often require the coordination of various sectors and government ministries, as climate change impacts are cross-cutting and affect multiple domains. However, in many countries of the Global South, there is a lack of effective inter-sectoral coordination, which hampers the implementation of integrated climate resilience strategies (Khalid, 2023; Nemakonde & Van Niekerk, 2023). In the fisheries, agriculture, pastoralist, and forest sectors, climate resilience is inherently multidisciplinary. For instance, addressing the impacts of climate change on agriculture requires coordination between the ministries of agriculture, water resources, and finance, among others. Similarly, forest management policies cannot succeed without the cooperation of forestry, land use, and environmental ministries. However, these ministries often operate in silos, with little collaboration between them. This fragmented approach leads to inefficiencies, duplication of efforts, and a lack of coherence in policy design and implementation. In the pastoralist sector, the challenge is even more pronounced, as climate resilience requires cooperation across sectors such as livestock management, water access, and land tenure (Rasul & Neupane, 2021). Pastoralists often move across borders and regions, complicating policy implementation and requiring cross-border coordination between national governments. The absence of such coordination exacerbates vulnerabilities and undermines resilience-building efforts. Moreover, the lack of institutional capacity to manage complex, cross-sectoral initiatives contributes to the failure of integrated approaches. Policymakers may not have the necessary technical knowledge or the resources to coordinate effectively across sectors, further undermining climate resilience efforts.

Weak institutional capacities in the Global South further impede the implementation of climate resilience policies. Many countries lack the necessary infrastructure, technical expertise, and human resources to design, implement, and monitor climate resilience initiatives effectively. This is particularly evident in the fisheries, agriculture, pastoralist, and forest sectors, where complex, locally tailored interventions are required to address the specific needs of vulnerable communities. For example, the fisheries sector often lacks the data and monitoring systems needed to track fish stocks and assess the impacts of climate change on marine ecosystems (Bolin et al., 2021). Similarly, in agriculture, there is a shortage of agricultural extension services and training programs to support farmers in adapting to changing climatic conditions (Antwi-Agyei & Stringer, 2021; Osumba et al., 2021). Pastoralist communities face challenges in accessing relevant climate information, and institutions responsible for managing grazing lands often lack the capacity to implement effective land-use policies (Senda et al., 2022). In the forest sector, weak enforcement of forest protection laws and inadequate capacity for monitoring illegal logging further exacerbate deforestation and forest degradation (Hifume et al., 2024). Without adequate institutional capacity, climate resilience policies cannot be effectively implemented or sustained. Building strong institutions that can support climate resilience requires long-term investment in human capital, infrastructure, and technical expertise, which many countries in the Global South struggle to provide.

Effective climate resilience policies require reliable data and a sound understanding of the local context. However, many countries in the Global South face significant knowledge gaps and data limitations, which hinder the formulation of appropriate policies (Aryal et al., 2021; Smidt & Jokonya, 2022). Climate data, such as accurate predictions of rainfall patterns, temperature changes, and the impacts of extreme events, is often sparse, unreliable, or outdated. In the fisheries and agriculture sectors, for instance, data on crop yields, water availability, and biodiversity are crucial for making informed decisions about adaptation measures. In addition, many communities in the Global South, particularly rural and remote areas, lack access to the information and training necessary to engage with climate resilience policies. This gap in knowledge and capacity can result in poor decision-making, as communities may not fully understand the risks they face or the options available for adaptation. Efforts to build local capacity for climate resilience require investments in education, information dissemination, and the integration of local knowledge with scientific data.

6.5 Future Climate Resilience Policy Directions in the Fishery, Agriculture, Pastoralist, and Forest Sectors in the Global South

Climate change presents a significant challenge to ecosystems, livelihoods, and economies across the Global South, particularly in the fishery, agriculture, pastoralist, and forest sectors. These sectors are not only highly dependent on natural resources but are also among the most vulnerable to climate variability and extreme weather events. To ensure long-term sustainability and resilience, it is crucial to develop climate resilience policies that are comprehensive, flexible, and inclusive. These policies must integrate sectoral needs while accounting for local realities and the complex socio-economic dynamics at play.

6.5.1 Integrating Sectoral Resilience Approaches

One of the key challenges in developing climate resilience policies is the need to recognize and address the interdependence between the fishery, agriculture, pastoralist, and forest sectors. Each of these sectors faces its own set of climate-related risks, but they also share common vulnerabilities, including water scarcity, changing rainfall patterns, temperature fluctuations, and increased incidences of pests and diseases. Therefore, any resilience policy must adopt a holistic, sectoral integration approach. This will involve promoting synergies across sectors and climate-smart agriculture (CSA) and fisheries. Cross-sectoral integration is essential for addressing shared vulnerabilities. For instance, policies that simultaneously support sustainable fisheries, agriculture, and pastoralism can have cumulative benefits. A policy that promotes agroforestry, which combines agricultural production with forest conservation, can increase both agricultural productivity and forest resilience. Similarly, implementing water management strategies that benefit both pastoralists and farmers such as watershed management, can ensure equitable access to water and enhance food and livelihood security. Climate-smart practices, which aim to increase productivity, enhance resilience, and reduce emissions, are critical in the agriculture and fishery sectors. In agriculture, CSA can include the promotion of drought-resistant crops, soil conservation techniques, and water-efficient irrigation systems. In fisheries, CSA can involve improving aquaculture practices, restoring coastal ecosystems like mangroves, and introducing adaptive management for fish stocks. Policy frameworks must be designed to scale these practices and ensure they are accessible to smallholder farmers and fisherfolk.

6.5.2 *Building Flexibility into Policy Frameworks*

Climate change impacts are inherently unpredictable, which calls for flexible policy frameworks that can adapt to evolving conditions. Flexibility in climate resilience policies is essential for responding to new risks, unexpected shocks, and emerging opportunities. This will involve the putting in place of adaptive governance structures, the use of climate information and early warning systems, as well as supporting innovation and technology transfer. Policymaking in the Global South must incorporate adaptive governance structures that allow for continuous learning and policy adjustments. This can be achieved through the establishment of multi-level governance systems, where local, regional, and national actors collaborate to assess climate risks and adjust policies accordingly. Local communities should have a central role in decision-making processes to ensure that policies remain grounded in the realities on the ground. Policies must be built on sound, data-driven decision-making processes. In agriculture, pastoralism, and fisheries, early warning systems are crucial for enabling timely responses to extreme weather events like floods, droughts, or cyclones. Governments and international organizations should invest in climate data infrastructure to enhance the availability and accuracy of forecasts, which can help sectors prepare in advance. Flexible policies should allow for rapid responses to these warnings, such as crop insurance for farmers or drought-resistant fodder for pastoralists. Climate resilience policies must also support innovation and the transfer of technologies. In the agriculture and fisheries sectors, innovative solutions such as drought-resistant crops, precision agriculture tools, and fishery management technologies, are crucial to building long-term resilience. However, these solutions must be accessible and appropriate for local contexts. Policymakers should facilitate the transfer of technologies and knowledge, particularly in rural areas, ensuring that local populations benefit from global advancements in climate resilience.

6.5.3 Inclusive Policies That Address Local Realities

An inclusive approach is fundamental to ensuring that climate resilience policies do not leave vulnerable populations behind. This includes marginalized communities such as women, indigenous peoples, smallholder farmers, and landless laborers, who often face heightened risks from climate change. Policies must recognize local realities, empowering these groups and providing them with the tools and resources to adapt to climate change. This should involve incorporating indigenous knowledge systems, gender-sensitive resilience policies as well as supporting rural communities and smallholders. Indigenous peoples, particularly in forested regions, possess invaluable knowledge about managing ecosystems and responding to environmental changes. This traditional knowledge, which has been passed down over generations, is often highly attuned to local climate conditions. Policies should seek to incorporate indigenous knowledge into climate resilience strategies, fostering partnerships between indigenous communities, governments, and researchers. This could include creating climate adaptation plans that blend modern science with traditional ecological practices. Gender is a critical dimension in climate resilience. In many regions of the Global South, women play central roles in agriculture, fisheries, and pastoralism, yet they often lack access to resources, decision-making processes, and climate-related information. Climate resilience policies must be gender-sensitive, ensuring that women have equal access to training, financial resources, and technology. Policies should also address the specific vulnerabilities women face, such as increased caregiving responsibilities during droughts or floods. Smallholder farmers and pastoralists are often the most vulnerable to climate change but are also the least equipped to manage these risks. Policies must provide direct support to these groups through access to credit, insurance, training, and market linkages. Additionally, land tenure security is crucial to enabling long-term investment in climate resilience. Without secure land rights, farmers and pastoralists may be reluctant to adopt climate-smart practices or improve their land management practices.

6.5.4 *Leveraging Financial and Market Instruments*

To support climate resilience, financial resources must be made available for adaptation initiatives. Governments, development organizations, and the private sector have important roles to play in mobilizing resources for climate resilience. This should involve climate financing mechanisms, and market-based solutions. Access to climate finance is often a barrier for countries in the Global South, where resources are scarce, and the impacts of climate change are most pronounced. Climate resilience policies should prioritize access to both international climate finance and domestic funding, through mechanisms such as the Green Climate Fund, national climate adaptation funds, and private-sector investments. These resources can support infrastructure development, capacity-building programs, and technology transfer. Market-based solutions can also play a role in enhancing resilience. For instance, farmers and fishers could be encouraged to adopt sustainable practices through market incentives, such as certification schemes for climate-resilient products or carbon credits. These market-based approaches could link sustainable practices in agriculture, forestry, and fisheries to global supply chains, thereby promoting economic growth alongside environmental protection.

References

AAI. (2024). *African adaptation initiative*. https://africaadaptationinitiative.org/

Aduhene-Chinbuah, J., Peprah, C. O., & cbxhgdfhd dZd s. (2024). Multi-risk management in Ghana's agricultural sector: Strategies, actors, and conceptual shifts—A review. *Review of Agricultural, Food and Environmental Studies, 105* 1–26.

Ahmadzai, M. R., Zaki, P. H., Ismail, M. H., Bawon, P., & Karam, D. S. (2022). The societal and economic impact of reforestation strategies and policies in Southeast Asia—A review. *Forests, 14*(1), 1.

Antwi-Agyei, P., & Stringer, L. C. (2021). Improving the effectiveness of agricultural extension services in supporting farmers to adapt to climate change:

Insights from northeastern Ghana. *Climate Risk Management, 32*, Article 100304.

Aryal, K., Laudari, H. K., Neupane, P. R., & Maraseni, T. (2021). Who shapes the environmental policy in the global south? Unpacking the reality of Nepal. *Environmental Science & Policy, 121*, 78–88.

Baskent, E. Z. (2024). A thorough assessment of various forest management planning initiatives and development of improvement strategies towards an ecosystem-based planning. *Environmental Development, 50*, Article 101006.

Bolin, J. A., Schoeman, D. S., Evans, K. J., Cummins, S. F., & Scales, K. L. (2021). Achieving sustainable and climate-resilient fisheries requires marine ecosystem forecasts to include fish condition. *Fish and Fisheries, 22*(5), 1067–1084.

Bryndum-Buchholz, A., Tittensor, D. P., & Lotze, H. K. (2021). The status of climate change adaptation in fisheries management: Policy, legislation and implementation. *Fish and Fisheries, 22*(6), 1248–1273.

Campbell, T. (2021). *Climate change policy narratives and pastoralism in the horn of Africa: New concerns, old arguments?* (Doctoral dissertation, Dublin City University).

Chichinye, A., Mlambo, D., Sebata, A., & Mabidi, A. (2024). Policy regarding sustainable forest management and resources. In *Agroforestry for carbon and ecosystem management* (pp. 403–417). Academic Press.

Cinner, J. E., Caldwell, I. R., Thiault, L., Ben, J., Blanchard, J. L., Coll, M., Diedrich, A., Eddy, T. D., Everett, J. D., Folberth, C., Gascuel, D., Guiet, J., Gurney, G. G., Heneghan, R. F., Jägermeyr, J., Jiddawi, N., Lahari, R., Kuange, J., Liu, W., & Pollnac, R. (2022). Potential impacts of climate change on agriculture and fisheries production in 72 tropical coastal communities. *Nature Communications, 13*(1), 3530.

Eschen, R., Bekele, K., Jumanne, Y., Kibet, S., Makale, F., Mbwambo, J. R., Megersa, B., Mijay, M., Moyo, F., Munishi, L., Mwihomeke, M., Nunda, W., Nyangito, M., Witt, A., & Schaffner, U. (2023). Experimental prosopis management practices and grassland restoration in three Eastern African countries. *CABI Agriculture and Bioscience, 4*(1), 21.

Fache, E., & Breckwoldt, A. (2018). Small-scale managed marine areas over time: Developments and challenges in a local Fijian reef fishery. *Journal of Environmental Management, 220*, 253–265.

FAO. (2025). *Climate-smart agriculture*. https://www.fao.org/climate-smart-agriculture/overview/en/

Frimpong, F., Asante, M., Peprah, C., Amankwaa-Yeboah, P., Danquah, E., Ribeiro, P., Aidoo, A., Agyeman, K., Asante, M., Keteku, A., & Botey, H. (2023). Water-smart farming: Review of strategies, technologies, and practices for sustainable agricultural water management in a changing climate in West Africa. *Frontiers in Sustainable Food Systems, 7*, 1110179.

Gakou Kakeu, J. V. (2022). *REDD+ integration, implementation, and interaction in the Congo basin: Evidence from Cameroon* (Doctoral dissertation, University of Leeds).

Hifume, A. K., Arias, J. M., & Anthony, O. (2024). Tackling Illegal logging: Problems and challenges. *American Journal of Humanities and Social Sciences Research (AJHSSR), 8*(5), 81–94.

Hussein, A. (2024). Climate smart agriculture strategies for enhanced agricultural resilience and food security under a changing climate in Ethiopia. *Sustainable Environment, 10*(1), 2345433.

IFAD. (2019). *Ethiopia: Pastoral community development project III*. https://www.ifad.org/en/w/projects/1100001522

ILRI. (2025). *International livestock research institute (ILRI)*. https://www.cgiar.org/research/center/ilri/

IPCC. (2022). Summary for Policymakers [H.-O. Pörtner, D.C. Roberts, E.S. Poloczanska, K. Mintenbeck, M. Tignor, A. Alegría, M. Craig, S. Langsdorf, S. Löschke, V. Möller, A. Okem, Eds.]. In *Climate change 2022: Impacts, adaptation and Vulnerability. Contribution of Working Group II to the Sixth Assessment Report of the Intergovernmental Panel on Climate Change* [H.-O. Pörtner, D. C. Roberts, M. Tignor, E. S. Poloczanska, K. Mintenbeck, A. Alegría, M. Craig, S. Langsdorf, S. Löschke, V. Möller, A. Okem, B. Rama (Ed.)] (pp. 3–33). Cambridge University Press. https://doi.org/10.1017/9781009325844.001. https://www.ipcc.ch/report/ar6/wg2/downloads/report/IPCC_AR6_WGII_SummaryForPolicymakers.pdf

IUCN. (2022). *Ecosystem-based Adaptation and the successful implementation and achievement of the sustainable development goals*. https://iucn.org/resources/file/ecosystem-based-adaptation-and-successful-implementation-and-achievement-sustainable

Iwama, A. Y., Araos, F., Anbleyth-Evans, J., Marchezini, V., Ruiz-Luna, A., Ther-Ríos, F., & Perkins, P. E. (2021). Multiple knowledge systems and participatory actions in slow-onset effects of climate change: Insights and perspectives in Latin America and the Caribbean. *Current Opinion in Environmental Sustainability, 50*, 31–42.

Kakraliya, S. K., Jat, H. S., Singh, I., Gora, M. K., Kakraliya, M., Bijarniya, D., Sharma, P. C., & Jat, M. L. (2022). Energy and economic efficiency of climate-smart agriculture practices in a rice–wheat cropping system of India. *Scientific Reports, 12*(1), 8731.

Katznelson, D., Sohns, A., Kim, D., Roozee, E., Donner, W. R., Song, A. M., de Vries, J. R., Temby, O., & Hickey, G. M. (2025). Examining the presence and effects of coherence and fragmentation in the Gulf of Maine fishery management network. *Regional Environmental Change, 25*(1), 1–16.

Khalid, A. M. (2023). Creating synergies among the sustainable development goals and climate action: Insights from a developing economy. *Sustainability, 15*(17), 13137.

Koh, N., Pietarinen, N., Ville, A., Kengoum, F., Wong, G. Y., & Brockhaus, M. (2024, March 20). Can REDD finance compete with established and emerging land investments? The case of Mai-Ndombe, Democratic Republic of Congo. Available at SSRN: https://ssrn.com/abstract=4766035; https://doi.org/10.2139/ssrn.4766035

Laudari, H. K., Aryal, K., Maraseni, T., Pariyar, S., Pant, B., Bhattarai, S., Kaini T. R., Karki, G., & Marahattha, A. (2022). Sixty-five years of forest restoration in Nepal: Lessons learned and way forward. *Land Use Policy, 115*, Article 106033.

Lelenguyah, G. L. (2023). *Analysis of Nexus between climate variability, herd mobility and livestock disease incidences in the Rangelands of Northern Kenya* (Doctoral dissertation, University of Nairobi).

Lipper, L., Thornton, P., Campbell, B. et al. (2014). Climate-smart agriculture for food security. *Nature Climate Change* 4, 1068–1072 (2014). https://doi.org/10.1038/nclimate2437

Lucantoni, D., & Domarle, J. (2023). Unlocking food security and livelihoods: The transformative power of agroecology among vulnerable smallholder farmers in Kembata Tembaro, Ethiopia. *Agroecology and Sustainable Food Systems, 47*(9), 1341–1371.

Marengo, J. A., Galdos, M. V., Challinor, A., Cunha, A. P., Marin, F. R., Vianna, M. D. S., Santos Alvalá, R. C., Alves, L., Mares, O., & Bender, F. (2022). Drought in Northeast Brazil: A review of agricultural and policy adaptation options for food security. *Climate Resilience and Sustainability, 1*(1), Article e17.

Mulungu, K., & Kangogo, D. (2022). Striving to be resilient: The role of crop-poultry integrated system as a climate change adaptation strategy in semiarid eastern Kenya. *Heliyon, 8*(11), Article e11579.

Mungai, E. M., Ndiritu, S. W., & Da Silva, I. (2022). Unlocking climate finance potential and policy barriers—A case of renewable energy and energy efficiency in Sub-Saharan Africa. *Resources, Environment and Sustainability, 7*, Article 100043.

Murphy, J. T. (2024). Green economies in the Global South: Framings, experiences, and immanent challenges. In *Research handbook on the green economy* (pp. 279–297). Edward Elgar.

Muthee, K., Duguma, L., Wainaina, P., Minang, P., & Nzyoka, J. (2022). A review of global policy mechanisms designed for tropical forests conservation and climate risks management. *Frontiers in Forests and Global Change, 4*, Article 748170.

National Fisheries Policy of India. (2020). *National Fisheries Policy, 2020: Sixth draft for consideration 30 December 2020*. https://dof.gov.in/sites/default/files/2020-12/Policy_0.pdf

Nemakonde, L. D., & Van Niekerk, D. (2023). Enabling conditions for integrating government institutions for disaster risk reduction and climate change adaptation in the SADC region and beyond. *Risk, Hazards & Crisis in Public Policy, 14*(1), 6–26.

Nguyen, T. H., Sahin, O., & Howes, M. (2021). Climate change adaptation influences and barriers impacting the Asian agricultural industry. *Sustainability, 13*(13), 7346.

Nori, M. (2022). *Assessing the policy frame in pastoral areas of Sub-Saharan Africa (SSA)*. https://cadmus.eui.eu/bitstream/handle/1814/74314/RSC_PP_2022_03_FINAL.pdf?sequence=1&isAllowed=y

Nori, M., & Scoones, I. (2023). Rethinking policies for pastoralists—Governing the rangelands. *The Rangeland Journal, 45*(2), 53–66.

Ojhaa, B., Pokharela, A., Adhikaria, B., & Bhattab, S. (2021). Status of watershed and need of integrated approach for sustainable water resources management in Nepal. *Big Data in Water Resources Engineering (BDWRE), 2*(1), 5–11.

Osumba, J. J., Recha, J. W., & Oroma, G. W. (2021). Transforming agricultural extension service delivery through innovative bottom–up climate-resilient agribusiness farmer field schools. *Sustainability, 13*(7), 3938.

Postigo, J. C. (2021). Navigating capitalist expansion and climate change in pastoral social-ecological systems: Impacts, vulnerability and decision-making. *Current Opinion in Environmental Sustainability, 52*, 68–74.

Purohit, S., Khan, Y. I., Hande, O. M., & Krishanmurti, & D'Souza, M. (2024). Green livelihoods as a nature-based solution for climate change mitigation and adaptation. *Handbook of nature-based solutions to mitigation and adaptation to climate change* (pp. 1–21). Springer.

Rasul, G., & Neupane, N. (2021). Improving policy coordination across the water, energy, and food, sectors in South Asia: A framework. *Frontiers in Sustainable Food Systems, 5*, Article 602475.

Resnick, D. (2024). Political economy of agriculture and food policy in Africa. *Annual Review of Resource Economics, 16*, 255–276.

Sacherer, A. K., Hoch, S., Dalfiume, S., & Kassaye, R. (2022). Financing forest conservation and restoration through climate policy instruments: lessons from the CDM and REDD+. In *Handbook of international climate finance* (pp. 293–317). Edward Elgar.

Selvaraj, J. J., Guerrero, D., Cifuentes-Ossa, M. A., & Alvis, Á. I. G. (2022). The economic vulnerability of fishing households to climate change in the south Pacific region of Colombia. *Heliyon, 8*(5), Article e09425.

Senda, T. S., Robinson, L. W., Gachene, C. K., & Kironchi, G. (2022). Formalization of communal land tenure and expectations for pastoralist livelihoods. *Land Use Policy, 114*, Article 105961.

Singh, S., Bhat, J. A., Shah, S., & Pala, N. A. (2021). Coastal resource management and tourism development in Fiji Islands: A conservation challenge. *Environment, Development and Sustainability, 23*(3), 3009–3027.

Smidt, H. J., & Jokonya, O. (2022). Factors affecting digital technology adoption by small-scale farmers in agriculture value chains (AVCs) in South Africa. *Information Technology for Development, 28*(3), 558–584.

UNDP. (2025). *Africa adaptation initiative (AAI).* https://www.adaptation-undp.org/sites/default/files/uploaded-images/aai_cas_launch.pdf

UNEP. (2025a). Building resilience to disasters and conflicts. https://www.unep.org/regions/africa/regional-initiatives/building-resilience-disasters-and-conflicts

UNEP. (2025b). *Green economy.* https://www.unep.org/pt-br/node/23750#:~:text=The%20UN%20Environment%20Programme%20has,in%20carbon%2C%20resource%20efficient%20and

Wando, D. O. (2023). *Environmental security and governance in resilience building for Eastern Mau forest communities, Nakuru County, Kenya* (Doctoral dissertation, Kenyatta University).

White, C. M., Mangubhai, S., Rumetna, L., & Brooks, C. M. (2022). The bridging role of non-governmental organizations in the planning, adoption, and management of the marine protected area network in Raja Ampat, Indonesia. *Marine Policy, 141*, Article 105095.

World Bank (2025). *Forests and landscapes: Healthy forests and landscapes support the foundations on which humans, wildlife, and ecosystems can thrive.* https://www.worldbank.org/en/topic/forests

Zerssa, G., Feyssa, D., Kim, D. G., & Eichler-Löbermann, B. (2021). Challenges of smallholder farming in Ethiopia and opportunities by adopting climate-smart agriculture. *Agriculture, 11*(3), 192.

CHAPTER 7

Conclusion to building resilience: climate change and livelihoods in the Global South

Abstract This chapter provides a comprehensive conclusion to the book, synthesizing the key findings on climate resilience and livelihoods in the Global South. It recaps strategies for resilience-building, community adaptations, and policy frameworks highlighted throughout the chapters. The reflection emphasizes the broader implications of these case studies, underscoring the importance of sustainable development in vulnerable regions. It discusses the crucial need for international cooperation and local action to strengthen climate resilience, recognizing that a collaborative approach is essential for addressing the challenges posed by climate change. The chapter also identifies priority areas for future research and outlines actionable steps for policymakers, NGOs, and local communities to enhance adaptive capacities. It concludes with a powerful call to action, urging stronger international commitments, the integration of indigenous knowledge systems, and the development of inclusive policies that support long-term resilience. Through this synthesis, the chapter highlights the importance of continuous, collective efforts in building climate resilience, offering a path forward for fostering sustainable livelihoods in the face of climate change.

Keywords Climate resilience · Global South · Adaptation · International cooperation · Local action · Sustainable development · Indigenous knowledge

N. P. Awazi, *Building Resilience*, Palgrave Studies in Climate Resilient Societies, https://doi.org/10.1007/978-3-031-93791-0_7

7.1 Summary of Key Findings

The Global South, encompassing regions that are most vulnerable to climate change, faces significant challenges in addressing the threats to ecosystems and livelihoods. As these areas experience rising temperatures, erratic weather patterns, and frequent extreme weather events, their populations—often dependent on natural resources for survival, are disproportionately affected. The book "Building Resilience: Climate Change and Livelihoods in the Global South" addresses these concerns across a variety of sectors, focusing on the implications for forest-dependent, fishing, pastoralist, and agriculture-dependent communities. The overarching theme throughout the chapters is the critical need for adaptive resilience strategies to reduce the vulnerability of these communities and enhance their long-term sustainability.

The introduction lays the groundwork by exploring the intersection between climate change and the vulnerabilities faced by the Global South. As climate impacts intensify, particularly for communities reliant on agriculture, forestry, fisheries, and pastoralism, the need for resilience becomes apparent. The chapter highlights how climate resilience refers to the capacity of individuals and ecosystems to absorb and recover from climate shocks. Central to building resilience is the integration of climate-smart practices, diversifying livelihoods, improving resource management, and providing communities with the knowledge and tools to confront climate impacts. The chapter underscores that strengthening adaptive capacities and ensuring marginalized communities are included in resilience efforts are essential to sustaining these livelihoods in the face of mounting climate threats.

Chapter 2 lays emphasis on forest-dependent communities and climate resilience. Forests are integral to the livelihoods of millions in the Global South. These forests provide food, income, and ecosystem services such as water purification and carbon sequestration. The chapter highlights the vulnerability of forest-dependent communities to climate change impacts such as altered rainfall patterns, pests, and droughts, all of which disrupt forest ecosystems and the services they provide. The findings stress the importance of local knowledge in shaping adaptive strategies, as these communities possess invaluable expertise in sustainable resource management. Moreover, the chapter emphasizes the potential of agroforestry, sustainable logging, and community-based conservation programs as ways to bolster resilience. These strategies promote environmental sustainability while ensuring that communities can adapt to changing conditions. The chapter calls for a policy framework that includes local knowledge and prioritizes the empowerment of these communities, ensuring both livelihood security and ecosystem conservation.

Chapter 3 focuses on fishing communities and adaptive management. Fishing communities in the Global South are acutely affected by climate change, particularly due to ocean acidification, rising sea temperatures, and shifting fish migration patterns. These environmental shifts threaten the core of many coastal livelihoods, which depend on stable fish stocks. This chapter explores adaptive management strategies such as diversifying livelihoods through aquaculture, ecotourism, and agro-fisheries, which can reduce vulnerability to climate risks. Traditional Ecological Knowledge (TEK), often rooted in centuries of experience, plays a crucial role in community-based resource management and conservation. The chapter underscores that integrating TEK with scientific approaches can help strengthen resilience. Policy frameworks at both the national and international levels are reviewed, with the key finding that these frameworks often fail to sufficiently support local fisheries. The chapter calls for stronger governance, increased community participation, and policies that ensure equitable access to resources for fishing communities, while also promoting sustainable practices.

Chapter 4 looks at pastoralist livelihoods in the face of climate change. Pastoralists in the Global South are among the most vulnerable groups in the context of climate change. Their livelihoods, which depend on grazing animals, are deeply impacted by climate-induced droughts, land degradation, and the loss of grazing lands. This chapter discusses how traditional resilience practices, such as transhumance (seasonal movement of livestock), communal land management, and the preservation of indigenous knowledge, continue to provide the foundation for resilience in pastoral communities. The chapter highlights that these practices are adaptive strategies that allow pastoralists to mitigate climate risks. Technological innovations, such as mobile weather applications and early warning systems, have also proven effective in enhancing resilience. However, the chapter notes that climate change continues to intensify the pressures on pastoral systems, demanding ongoing adaptation. It concludes that the future of pastoralism in the Global South hinges on diversifying income sources, accessing new markets, and securing policy support for sustainable land management and resource access.

Chapter 5 examines agricultural communities and climate resilience. Agriculture-dependent communities in the Global South face considerable threats from climate change, particularly in the form of unpredictable rainfall, floods, and prolonged droughts. These environmental changes threaten food security and undermine agricultural productivity. However, the chapter emphasizes that adaptive agricultural practices, such as crop diversification, agroforestry, and the use of climate-resilient crop varieties, have proven effective in mitigating some of these impacts. Yet, the capacity of these communities to fully embrace climate-smart practices is often constrained by a lack of access to resources, technology, and knowledge. Farmers in these regions often lack sufficient education and extension services, which hinders their ability to implement adaptive strategies. The chapter stresses that policy interventions must focus on improving infrastructure, finance, and rural development to facilitate the adoption of climate-resilient agricultural practices. It also emphasizes the importance of localized approaches, as these are more likely to resonate with the realities of farmers' lives. Successful case studies, where such practices have been adopted, demonstrate the potential for climate resilience in agricultural communities when supported by robust policy frameworks.

Chapter 6 assesses climate resilience policies across sectors. This chapter examines the state of climate resilience policies across the fisheries, agriculture, pastoralism, and forestry sectors in the Global South. It explores the challenges that hinder the effective implementation of these policies, including inadequate funding, political inertia, and lack of intersectoral coordination. The chapter presents ecosystem-based adaptation (EbA) and climate-smart agriculture (CSA) as innovative approaches that can contribute to sustainable development goals and promote green economies. These approaches are seen as complementary to traditional resilience strategies and can enhance the ability of communities to cope with climate change impacts. The chapter identifies the need for policies that are more inclusive, flexible, and able to accommodate the realities of local communities. It concludes with a call for integrated, cross-sectoral approaches that promote cooperation and recognize the vital role of local knowledge in designing effective resilience strategies.

Thus, building climate resilience in vulnerable sectors in the Global South requires a multifaceted approach that integrates local knowledge, strengthens adaptive capacity, and supports policy frameworks that are inclusive and community-driven. The case studies from forest, fishing, pastoralist, and agricultural communities demonstrate the importance of adaptive management strategies, diversification of livelihoods, and the implementation of climate-smart practices. The emphasis on community-led solutions and policy frameworks that support these practices is critical to reducing vulnerability and ensuring the long-term sustainability of livelihoods in the face of climate change. The book stresses on the fact that, while climate change poses significant challenges, these can be mitigated through proactive, community-based approaches that are supported by robust policies and investment in capacity-building. Climate resilience is not only about managing risks but also about empowering communities to take charge of their own futures, ensuring that they can withstand the shocks of today and adapt to the uncertainties of tomorrow. Therefore, resilience in the Global South depends on collaboration, innovation, and a commitment to addressing the root causes of vulnerability in the context of an ever-changing climate.

7.2 Reflections on Climate Resilience in the Global South

The Global South, which includes many of the world's poorest and most vulnerable regions, is increasingly at the forefront of climate change's adverse effects. These regions, which rely heavily on sectors such as agriculture, fishing, forestry, and pastoralism for their livelihoods, are facing profound challenges as climate change exacerbates existing vulnerabilities. The book *Building Resilience: Climate Change and Livelihoods in the Global South* sheds light on how these communities are coping with the impacts of climate change and highlights strategies that can foster long-term sustainability. By examining case studies from forest-dependent communities, fishing communities, pastoralist livelihoods, and agriculture-dependent populations, the book offers significant insights into the challenges and opportunities for climate change adaptation and sustainable development in vulnerable regions. Reflecting on these case studies provides a deeper understanding of the broader implications for climate resilience in the Global South and the path forward for sustainable development in these areas.

7.2.1 Vulnerability and the Need for Adaptation

The central theme throughout the book is the vulnerability of the Global South to climate change, particularly in regions where economic and social development is already constrained. The populations in these areas are often highly dependent on natural resources, such as fertile land, forests, and fisheries, to sustain their livelihoods. In turn, these resources are increasingly threatened by changing climate patterns, such as rising temperatures, altered rainfall distributions, and more frequent extreme weather events like droughts and floods. The case studies from the fisheries, forestry, pastoralist, and agriculture sectors show that the impacts of climate change are multifaceted and disproportionately affect marginalized communities that are already struggling with poverty, inequality, and lack of access to technological and financial resources. Climate-induced disruptions such as crop failures, the loss of grazing land, declining fish stocks, and deforestation place immense pressure on these communities. Adaptation to these changes is no longer optional but essential for the survival and long-term sustainability of these sectors. However, adaptation is not a one-size-fits-all solution. As the book illustrates, each community, sector, and region has unique needs and challenges that require tailored approaches. What emerges clearly from the case studies is that while the risks are universal, the solutions must be locally specific, incorporating both modern scientific knowledge and traditional, community-based practices. This dual approach can foster resilience by drawing on the strengths of both formal scientific research and indigenous knowledge systems, which have evolved over centuries of close interaction with local ecosystems.

7.2.2 Building Resilience Through Diversified Livelihoods

One of the most striking themes in the case studies is the importance of diversifying livelihoods as a key strategy for building climate resilience. For example, in fishing communities, the reliance on a single source of income—fishing is increasingly untenable as fish stocks decline due to overfishing and climate change. By diversifying into aquaculture, ecotourism, or agro-fisheries, these communities can spread risk and increase their economic security. Similarly, for pastoralists, diversifying income sources beyond livestock, such as through agricultural activities or trade, can help buffer the impacts of climate-induced droughts and land degradation. Agricultural communities face a similar need to diversify. In many regions, farmers depend on a few staple crops that are highly susceptible to climate variability, such as maize or rice. Adopting practices like crop diversification, agroforestry, and the use of climate-resilient crop varieties can help to reduce vulnerability. While diversification provides a way to cope with immediate shocks, it also strengthens communities' long-term resilience by creating a more stable and diverse economic base. However, the ability to diversify livelihoods is not always easy or accessible to all. As the book highlights, many of the communities in the Global South lack access to the necessary resources, knowledge, or technology to implement such strategies. Barriers such as limited access to markets, credit, or training programs can hinder the potential benefits of diversification. Therefore, it is imperative that climate resilience efforts focus on reducing these barriers and providing communities with the necessary tools to pursue alternative livelihoods.

7.2.3 The Role of Policy and Governance

An essential reflection on the case studies presented in the book is the central role of policy and governance in shaping the success of climate resilience efforts. Effective policies are crucial in enabling communities to adopt climate-smart practices and strategies for resilience. For example, the case studies from forest-dependent communities emphasize the need for policies that support sustainable forest management, including agroforestry and community-based conservation programs. These policies should not only address environmental concerns but also enhance socioeconomic stability by fostering the involvement of local communities in decision-making processes. In the fisheries sector, national and international policies that promote sustainable fisheries management and protect marine ecosystems can help prevent the overexploitation of resources and ensure that fishing communities can continue to rely on these ecosystems. Similarly, policies that facilitate access to new markets, technologies, and resources for pastoralists and farmers are critical in enhancing their resilience. However, the book also identifies key challenges in policy implementation, such as inadequate funding, political inertia, and a lack of coordination across sectors. These barriers can undermine the effectiveness of climate resilience measures, even when good policies exist on paper. The case studies underscore the importance of inclusive, flexible, and bottom-up policy frameworks that recognize the specific needs and knowledge of local communities. Policy interventions must be participatory, ensuring that marginalized groups have a voice in shaping the policies that affect their livelihoods. Additionally, the integration of climate resilience into national development strategies is essential for long-term success. Policies that are siloed or isolated from broader development goals are less likely to address the systemic challenges posed by climate change.

7.2.4 Ecosystem-Based Adaptation and Climate-Smart Approaches

The concept of ecosystem-based adaptation (EbA) emerges as a promising approach to enhancing climate resilience in the Global South. EbA emphasizes the role of ecosystems such as forests, wetlands, and coastal areas in providing essential services that help communities adapt to climate change. The case studies in the forestry and fishing sectors illustrate how ecosystem-based approaches can protect biodiversity, improve water availability, and enhance food security, all of which contribute to greater resilience. Similarly, the book discusses climate-smart agriculture (CSA) as an important strategy for building resilience in agriculture-dependent communities. CSA integrates climate change adaptation, mitigation, and sustainable development goals, promoting practices that enhance productivity while reducing environmental impacts. The case studies highlight how farmers have adopted CSA practices, such as using drought-resistant crop varieties and improving soil health, to cope with climate variability. Both EbA and CSA demonstrate the critical link between environmental sustainability and climate resilience. These approaches not only help communities adapt to climate change but also contribute to the broader goals of sustainable development by preserving ecosystems, improving livelihoods, and reducing the risk of future climate-related disasters. As such, they should be prioritized in climate resilience efforts.

7.2.5 A Path Forward for Sustainable Development

Ultimately, the reflections on the case studies point to the need for a holistic and integrated approach to climate change adaptation in the Global South. Climate resilience cannot be achieved through isolated efforts in individual sectors; rather, it requires coordinated action that combines environmental, social, and economic dimensions. It is crucial to recognize the interconnectedness of different sectors and communities, and to foster solutions that work across sectors to promote long-term sustainability. The book makes clear that climate resilience in the Global South is not just about survival in the face of climate impacts; it is about thriving and ensuring that vulnerable populations can adapt to and benefit from the changing climate. This can only be achieved by empowering local communities, integrating indigenous knowledge, and creating supportive policy environments that enable them to build their own resilience. As the Global South continues to bear the brunt of climate change, the lessons learned from the case studies in *Building Resilience* offer important insights into the strategies that can help communities weather the storm. By focusing on adaptive management, diversified livelihoods, inclusive policies, and ecosystem-based solutions, vulnerable regions can lay the foundation for a more resilient and sustainable future.

7.3 The Role of International Cooperation and Local Action

As climate change accelerates, the Global South faces some of the most severe challenges due to the direct impacts on ecosystems and livelihoods. Communities in these regions are heavily dependent on sectors such as agriculture, fisheries, forestry, and pastoralism, which are highly vulnerable to shifting climate patterns. Addressing these challenges requires an integrated approach that draws on both local action and international cooperation. Strengthening climate resilience in the Global South demands a collaborative effort, combining the expertise, resources, and capacities of both global and local stakeholders. While local action is essential to tailoring solutions to specific needs and contexts, international cooperation can provide the financial support, technological resources, and policy frameworks necessary to empower local communities. Together, these efforts can build the foundation for long-term resilience and sustainable development in vulnerable regions.

7.3.1 *The Need for a Collaborative Approach*

A key theme in the book *Building Resilience: Climate Change and Livelihoods in the Global South* is the recognition that climate resilience cannot be built in isolation. Vulnerable communities in the Global South are deeply affected by climate change, but they also possess valuable local knowledge, adaptive strategies, and resilience practices that can provide the foundation for effective climate adaptation. However, these communities often lack the resources, financial support, and technical expertise needed to scale these efforts or to access global markets, technology, or climate financing. International cooperation plays a critical role in providing these communities with access to these necessary resources. At the same time, global actors must understand that solutions to climate resilience must be locally specific. A top-down approach where international organizations or governments impose blanket solutions without considering the unique needs of local communities is unlikely to succeed. Instead, a collaborative approach that values the knowledge and priorities of local stakeholders, while also leveraging global expertise is essential. A collaborative approach involves ensuring that international and local actors work together in a way that is mutually beneficial. While the Global South has been historically marginalized in global governance discussions, it is crucial to bring local voices to the table when designing policies or funding initiatives aimed at strengthening resilience. This process requires building effective partnerships that combine the strengths of both levels of action.

7.3.2 The Role of International Cooperation

International cooperation is fundamental to addressing the climate resilience challenges faced by the Global South. One of the most important ways in which the global community can contribute is through the provision of financial support. Many countries in the Global South are not only the most vulnerable to climate change but also the least equipped to deal with its impacts. Financial resources are crucial for enabling communities to adapt and build resilience. International organizations, development banks, and donor countries play a key role in providing this financial support. Mechanisms like the Green Climate Fund (GCF) and the Adaptation Fund are designed to provide financing for adaptation projects in vulnerable regions. These funds can support initiatives that enhance agricultural productivity, improve water management systems, promote sustainable forestry practices, or diversify livelihoods, all of which can contribute to building resilience in the face of climate change. Additionally, international cooperation is vital for transferring technology and expertise. Climate change adaptation often requires the use of advanced technologies such as early warning systems, drought-resistant crops, or renewable energy solutions, that may not be readily accessible in the Global South. International partnerships can facilitate the transfer of these technologies to local communities, while also providing training and capacity-building to ensure they are used effectively. This technological support can help increase adaptive capacity and reduce vulnerability to climate impacts. Moreover, global frameworks and agreements, such as the Paris Agreement, create platforms for collaboration between countries on climate action. The Paris Agreement specifically emphasizes the need for developed countries to assist developing countries in their climate efforts, including through financing, capacity-building, and technology transfer. These international agreements provide a formalized mechanism for countries to commit to actions that enhance resilience, ensuring that the Global South is not left behind in the global response to climate change. However, the effectiveness of international cooperation depends on the commitment and accountability of all parties involved. While financial resources and technology transfer are critical, they must be accompanied by transparent, efficient implementation to ensure that funds are used appropriately and that communities benefit from these resources. Without strong governance and accountability mechanisms, international support can be misallocated or fail to reach the communities that need it most.

7.3.3 The Power of Local Action

While international cooperation provides essential support, local action is where climate resilience is built and sustained. Communities in the Global South often have a deep understanding of their local environments, having adapted over generations to environmental changes and challenges. This local knowledge is a vital resource that can help communities identify sustainable solutions to climate-related problems. Local knowledge systems are rooted in the cultural, historical, and environmental contexts of specific regions, allowing communities to develop adaptive strategies that reflect their unique challenges and opportunities. For example, in forest-dependent communities, indigenous practices such as sustainable logging, agroforestry, and community-led conservation programs have long been essential for maintaining the health of ecosystems and sustaining livelihoods. These traditional practices are often more sustainable and climate-resilient than industrial approaches, as they are designed to work with the environment rather than against it. Supporting local initiatives, such as these community-based conservation efforts, can enhance climate resilience while preserving biodiversity and improving livelihoods. In fishing communities, traditional ecological knowledge (TEK) plays a key role in managing marine resources sustainably. TEK includes local insights into fish migration patterns, seasonal variations in marine resources, and sustainable harvesting methods. These insights are critical in fostering resilience in the face of climate change, which is disrupting marine ecosystems through warming waters, ocean acidification, and shifting fish migration patterns. When combined with scientific approaches, TEK can lead to more sustainable and adaptive fisheries management systems. Moreover, local action is vital for ensuring that climate adaptation strategies are relevant, effective, and culturally appropriate. Communities often face different challenges depending on their geographic location, economic status, and reliance on specific natural resources. Local actors, including community leaders, farmers, fisherfolk, and pastoralists, have the knowledge and capacity to design and implement solutions that are tailored to their needs. These solutions can range from implementing climate-resilient farming practices, such as agroforestry, to creating local early warning systems or strengthening community-based disaster preparedness plans. However, local action often faces significant barriers, including limited access to resources, knowledge, and capacity. In many cases, local communities need support from external stakeholders—governments, NGOs, and international partners, to scale up their efforts. This is where the collaboration between local and international actors becomes crucial.

7.3.4 *Strengthening the Collaborative Approach*

To strengthen climate resilience in the Global South, a collaborative approach between international and local actors must go beyond just providing financial resources or technical assistance. Successful climate resilience programs should aim for long-term partnerships that empower local communities while supporting their capacity to engage with broader national and global processes. This collaboration should prioritize capacity-building and knowledge-sharing, inclusive decision-making, strengthening governance and accountability, as well as sustainability and long-term support. International cooperation should focus on building the capacity of local stakeholders, enabling them to design, implement, and sustain climate resilience strategies. This includes providing training in climate adaptation techniques, promoting knowledge-sharing networks, and fostering cross-sectoral collaboration between governments, civil society, and the private sector. Local communities should be involved in decision-making processes related to climate adaptation. International actors must support inclusive, participatory processes that ensure the voices of vulnerable communities are heard and incorporated into policy design. This approach strengthens ownership of adaptation strategies and ensures their relevance to local contexts. Effective governance mechanisms are critical for ensuring that both international and local actions are aligned with resilience objectives. Transparent and accountable management of resources, policies, and programs is essential to maximize the impact of climate resilience efforts. This includes ensuring that funds reach the intended communities and that projects are monitored and evaluated to assess their effectiveness. International cooperation should focus on providing not just short-term financial support but also long-term commitments that enable sustainable adaptation. This includes facilitating access to climate insurance, offering technical assistance for the long-term maintenance of adaptation strategies, and supporting community-led initiatives that are built to last.

7.4 Future Research and Action

As climate change continues to exacerbate vulnerabilities in the Global South, it becomes increasingly urgent for researchers, policymakers, NGOs, and local communities to collaborate and innovate solutions that enhance resilience. While progress has been made, the challenges posed by climate change are vast and multifaceted, requiring continued investment in research and action. Future research will play a critical role in refining adaptation strategies, while practical steps for policymakers, NGOs, and local communities will ensure that these solutions are scalable, inclusive, and effective. Key areas for future research should be:

7.4.1 Understanding Local Climate Variability and Ecosystem Changes

One of the critical gaps in current research is a more nuanced understanding of local climate variability and ecosystem changes at the micro-level. While much attention is given to broad regional and global climate patterns, there is a need for more localized, community-specific data to understand how climate change is manifesting in particular regions. This includes changes in weather patterns, soil health, water availability, and biodiversity shifts. Localized climate models can provide essential insights into how specific ecosystems are being impacted and how different communities experience climate change in unique ways. Research should focus on localized climate data collection through community-based monitoring systems; mapping climate impacts on local ecosystems and natural resource availability; and identifying region-specific vulnerabilities in sectors like agriculture, fisheries, pastoralism and forestry. Such localized data will enable communities to adapt more effectively by using the information to develop tailored resilience strategies that match their specific environmental and socio-economic conditions.

7.4.2 Integrating Indigenous Knowledge and Traditional Ecological Knowledge (TEK)

Indigenous communities in the Global South have developed a wealth of traditional ecological knowledge (TEK) over centuries, which has enabled them to live sustainably within their environments. However, this knowledge has often been overlooked or marginalized in favor of scientific approaches. Future research should explore ways to integrate TEK with contemporary climate science and technology to develop more adaptive and resilient solutions. Research in this area could include: documentation and validation of traditional knowledge practices and how they can complement modern climate adaptation strategies; studying the role of indigenous governance systems in managing natural resources and responding to climate challenges; and developing frameworks for combining scientific and indigenous knowledge to enhance resilience and sustainability. Through the blending indigenous of knowledge with modern scientific tools, communities can better address the complexities of climate adaptation and ensure that resilience strategies respect cultural heritage and local practices.

7.4.3 Assessing the Effectiveness of Climate Adaptation Practices

There is a growing body of knowledge on climate adaptation practices, such as agroforestry, crop diversification, and sustainable fisheries management. However, there is a need for more in-depth, long-term studies to evaluate the effectiveness of these practices in enhancing resilience. Future research should focus on monitoring the outcomes of existing adaptation projects to identify best practices and scale successful models. Research should aim to conduct longitudinal studies on the outcomes of climate adaptation interventions in agriculture, fisheries, and pastoralism; analyze the socio-economic impacts of resilience-building strategies on marginalized groups, such as women, youth, and indigenous peoples; and explore the barriers to implementing successful adaptation practices, including access to finance, knowledge, and technology. Through an understanding of what works and why, future adaptation efforts can be refined and optimized to meet the unique challenges faced by different communities.

7.4.4 *Innovative Financing Mechanisms for Climate Resilience*

One of the most significant challenges to building resilience in the Global South is the lack of financing for adaptation projects. While international funding mechanisms such as the Green Climate Fund (GCF) exist, they often face issues of inefficiency, complexity, and unequal access. Future research should focus on exploring new and innovative financing mechanisms that can ensure adequate and equitable funding for climate resilience efforts. Research in this area could explore: the potential of microfinance and social impact investing to support local adaptation projects; innovative financing models such as blended finance, which combines public and private sector funding to scale up resilience efforts; the role of insurance and risk financing instruments, such as climate insurance, in reducing vulnerability. Exploring diverse and innovative financing mechanisms is essential for creating sustainable funding models that can be scaled across different sectors and regions.

For policymakers, NGOs, and local communities, the following practical steps should be followed:

7.4.5 *Developing Localized and Inclusive Policies*

Policymakers in the Global South need to focus on creating policies that are both inclusive and localized, drawing from community needs and priorities. National and regional climate resilience strategies should integrate the voices of local stakeholders, especially marginalized and vulnerable groups, such as women, youth, and indigenous communities. Policies should also be flexible enough to accommodate the diverse challenges faced by different regions within the same country. Practical steps include strengthening participatory governance mechanisms where local communities are actively involved in policy design and implementation;

supporting policy frameworks that integrate climate change adaptation into broader development plans, such as agriculture, education, and healthcare; promoting policies that enable the transfer of technology and resources to local communities, particularly in sectors like agriculture and fisheries. These steps will ensure that resilience-building strategies are not only effective but also equitable, empowering communities to take charge of their own adaptation efforts.

7.4.6 Scaling Up Community-Based Adaptation Programs

NGOs and international organizations can play a crucial role in scaling up community-based adaptation programs. While many grassroots-level initiatives have shown success, they often lack the resources or political support to expand. NGOs should work with governments to help local communities access the tools and resources needed to scale successful resilience projects. Actionable steps include strengthening partnerships between NGOs, governments, and local communities to foster knowledge exchange and joint action; securing long-term funding commitments for community-led adaptation projects; providing training and capacity-building programs that enhance the ability of local communities to implement and manage climate adaptation projects. Through a scaling up of successful community-based initiatives, it is possible to create broader systemic changes that enhance climate resilience across regions.

7.4.7 Building Capacity for Climate Adaptation

Both local communities and national governments need to develop greater capacity for climate adaptation. This involves providing local actors with the tools, knowledge, and skills to cope with climate impacts. Training programs should focus on a variety of areas, including disaster risk reduction, sustainable resource management, and the adoption of climate-resilient agricultural practices. Practical steps include: organizing local training workshops and capacity-building initiatives that help communities understand and implement climate adaptation strategies; strengthening local technical expertise in climate science, agriculture, water management, and environmental governance; and promoting cross-sectoral collaboration and knowledge-sharing among local, national, and international stakeholders. This capacity-building will ensure that communities are not only prepared for climate impacts but also equipped to take proactive steps in building resilience.

7.4.8 *Leveraging Technology for Adaptation*

Technology plays an essential role in enhancing climate resilience, particularly in areas such as early warning systems, water management, and climate-smart agriculture. Local communities must have access to the technologies that can improve their capacity to adapt to climate change. Policymakers and NGOs should work together to facilitate the transfer and appropriate use of technology in the Global South. Actionable steps include: supporting the development of low-cost, accessible technologies that can be easily adopted by local communities; expanding access to mobile technology that can provide farmers, pastoralists, and fishermen with real-time climate information and market data; and promoting innovations in renewable energy, water management, and crop resilience through public–private partnerships.

7.5 Call to Action

The impacts of climate change are being felt most acutely in the Global South, where vulnerable communities face disproportionate risks due to their reliance on climate-sensitive sectors like agriculture, fisheries, forestry, and pastoralism. The time for action is now. We must prioritize stronger international commitments, better integration of indigenous knowledge, and more inclusive policies to build long-term climate resilience in these regions. Only through collective action, collaboration, and a shift toward more inclusive and equitable strategies can we hope to secure a resilient future for the most vulnerable populations on our planet.

7.5.1 *Stronger International Commitments*

International cooperation remains one of the most critical components in addressing climate change in the Global South. While many countries and organizations have made commitments to tackle climate change through agreements like the Paris Accord, these efforts must be further strengthened. Developed countries, which are historically responsible for the majority of greenhouse gas emissions, must fulfill their financial commitments to support climate adaptation and resilience efforts in developing countries. International actors must work to increase funding through mechanisms like the Green Climate Fund, ensuring that financial resources reach those who need them most. Additionally, it is essential that international cooperation extends beyond financial aid to include knowledge transfer, technology, and capacity-building initiatives that can empower local communities. This includes providing access to renewable energy solutions, sustainable agricultural technologies, and climate-resilient infrastructure. Without these tools, local communities will struggle to effectively adapt to climate change, undermining resilience-building efforts. There should be more advocacy for stronger commitments to policy coherence at the global level. Climate change is a complex, multifaceted issue that requires coordinated action across various sectors such as agriculture, water, health, and more. Governments, NGOs, and international organizations must ensure that climate resilience is integrated into all aspects of development policy. Cross-sectoral cooperation is vital for creating strategies that address both the immediate impacts of climate change and the longer-term needs of vulnerable populations.

7.5.2 *Better Integration of Indigenous Knowledge*

Indigenous peoples in the Global South possess valuable Traditional Ecological Knowledge (TEK) that has been cultivated over generations. This knowledge has allowed communities to adapt to changing environments in sustainable ways. Despite its importance, indigenous knowledge has often been disregarded in favor of top-down scientific approaches that do not always consider local contexts. It is time to acknowledge the critical role that indigenous knowledge plays in building resilience. Indigenous practices in land management, agriculture, and resource conservation have long demonstrated the ability to foster environmental sustainability and adapt to climate stress. By integrating indigenous knowledge into formal climate adaptation strategies, we can ensure that solutions are both contextually relevant and culturally appropriate. Collaborative research between scientists and indigenous communities should be promoted to develop innovative, hybrid solutions that blend modern scientific understanding with traditional wisdom. Policymakers, NGOs, and international organizations must recognize the value of indigenous governance systems and promote community-led decision-making processes. This will not only enhance the effectiveness of climate resilience efforts but also empower local communities to take ownership of their futures.

7.5.3 More Inclusive Policies

Lastly, climate resilience policies must be inclusive. Vulnerable groups, particularly women, youth, and marginalized communities, often face compounded risks from climate change, and their voices must be included in the policy-making process. Inclusive policies are necessary to ensure that adaptation strategies address the specific needs of these groups and offer equitable solutions that leave no one behind. This involves ensuring that climate adaptation strategies are not one-size-fits-all but instead reflect the unique needs of different communities. Policymakers must work closely with local communities to co-design policies that take into account local contexts, values, and priorities. Furthermore, it is essential to address the structural barriers such as poverty, gender inequality, and lack of access to resources—that exacerbate vulnerability and limit adaptive capacity. International and national governments should prioritize gender-responsive climate policies, providing women with access to resources, land rights, and decision-making processes. Youth engagement is also crucial in shaping the future of resilience-building efforts. Empowering the next generation to become climate leaders will ensure that resilience strategies are not only sustained but also continuously adapted as the climate crisis evolves.

Index

N. P. Awazi, *Building Resilience*, Palgrave Studies in Climate Resilient Societies, https://doi.org/10.1007/978-3-031-93791-0

Y

www.ingramcontent.com/pod-product-compliance
Lightning Source LLC
Chambersburg PA
CBHW070041210725
29875CB00004B/52
* 9 7 8 3 0 3 1 9 3 7 9 0 3 *